Springer
Berlin
Heidelberg
New York
Barcelona
Budapest
Hong Kong
London
Milan
Paris
Santa Clara
Singapore
Tokyo

Francisco J. Barrantes (Ed.)

The Nicotinic Acetylcholine Receptor: Current Views and Future Trends

Springer

Francisco J. Barrantes

Instituto de Investigaciones
Bioquímicas de Bahía Blanca
Universidad Nacional del Sur
Bahía Blanca, Argentina

ISBN: 3-540-64258-7 Springer-Verlag Berlin Heidelberg New York
Biotechnology Intelligence Unit

Library of Congress Cataloging-in-Publication data

The nicotinic acetylcholine receptor : current views and future trends /
[edited by] Francisco Jose Barrantes.
p. cm.--(Biotechnology intelligence unit)
Includes bibliographical references and index.
ISBN (invalid) 1-57059-414-3 (alk. paper)
1. Nicotinic receptors. 2. Acetylcholine--Receptors.
I. Barrantes, Francisco Jose, 1944- . II. Series.
[DNLM: 1. Receptors, Nicotinic--physiology. 2. Receptors. Cholinergic--physiology. WL 102.8 N6618 1997]
QP364.7.N54 1997
612.8'14--dc21
DNLM/DLC
for Library of Congress 97-36942
CIP

Typesetting: R.G. Landes Company, Georgetown, TX, U.S.A.

SPIN 10674071 31/3111 - 5 4 3 2 1 0 - Printed on acid-free paper

Dedication

To Phyllis

FOREWORD

Over the last two decades a convergence of techniques from various scientific disciplines has contributed to our comprehension of the structure, evolutionary trends and the multiplicity of functions performed by ligand- and voltage-gated ion channels and receptors. This book puts together in a well organized, comprehensive and yet succinct manner one of the fastest growing fields in the Molecular Neurosciences, that of the ligand-gated ion channels (LGIC). Several neurotransmitter receptors constitute this superfamily, the nicotinic acetylcholine receptor still being the prototype.

The study of central nervous system (neuronal) acetylcholine receptors is generating growing interest owing to their likely involvement in nicotine addiction and other pathological conditions, and the possibility of developing pharmacological compounds exploiting the positive effects of nicotine (anxiolysis, anti-depression, cognitive enhancement) without the negative health consequences of tobacco usage. Future trends are analyzed in each chapter; they encompass the likely strategies to be employed in decoding structure-function relationships of the receptor molecule, in establishing the differences in ionic selectivity and establishing the mechanisms of permeability through the pore, and the possible use of genetic kindredness between acetylcholine receptors of different species for the diagnosis of new genetic diseases.

CONTENTS

EDITOR

Francisco J. Barrantes
Instituto de Investigaciones Bioquímicas de Bahía Blanca
Universidad Nacional del Sur
Bahía Blanca, Argentina
Chapters 1, 5, 8

CONTRIBUTORS

Georgina E. Barrantes
Instituto de Neurociencias
Departamento de Biología
Buenos Aires, Argentina *and*
Instituto de Investigaciones Bioquímicas de Bahía Blanca
Bahía Blanca, Argentina
Chapter 5

Meyer B. Jackson
Department of Physiology
University of Wisconsin - Madison
Madison, Wisconsin, U.S.A.
Chapter 4

Ronald J. Lukas
Division of Neurobiology
Barrow Neurological Institute
Phoenix, Arizona, U.S.A.
Chapter 7

Marcelo O. Ortells
Instituto de Neurociencia
Facultad de Ciencias Exactas y Naturales
Universidad de Buenos Aires, Buenos Aires Argentina *and*
Instituto de Investigaciones Bioquímicas de Bahía Blanca
Bahía Blanca, Argentina
Chapters 2, 5

Richard J. Prince
Receptor Biology Laboratory
Department of Physiology and Biophysics
Mayo Foundation
Rochester, Minnesota, U.S.A.
Chapter 3

Steven M. Sine
Receptor Biology Laboratory
Department of Physiology and Biophysics
Mayo Foundation
Rochester, Minnesota, U.S.A.
Chapter 3

Alfredo Villarroel
Department of Physiology and Biophysics
Dalhousie University School of Medicine
Halifax, Nova Scotia, Canada
Chapter 6

CHAPTER 1

Introduction: Structure Meets Function at the Acetylcholine Receptor

Francisco J. Barrantes

When is a topic mature for critical analysis? Has the field of the nicotinic acetylcholine receptor (AChR) reached the necessary stage of "ripeness" to warrant such an exercise? Any nonspecialized readers following its development by browsing through reviews on the subject over the last two decades may have gotten the impression that the topic has "...always been *almost* finished; only the detailed high-resolution structure is still lacking...." But as a Spanish philosopher put it, "there are no exhausted subjects, but men exhausted in the pursuance of such subjects...." The more we learn about the AChR, the more questions that can be formulated, and the more complex the subject becomes.

What has made the AChR field unique for almost two decades is that it has provided us with a perfect object to contemplate a single molecule in action, thanks to its inherent ability to amplify the signal stemming from the passage of thousands of ions through its intrinsic pore, the channel. Before this was experimentally feasible using the technique we know as patch-clamp recording, biochemists had already taken a first glance at the AChR through isolation, purification and characterization procedures. In this reductionist approach—biochemistry is, after all, a form of molecular dissection (or molecular "anatomy" from the Greek ανα′ = up; τομ = cut; Caelius Aurelianus ca. 420)—pounding the appropriate tissue and having the right pharmacological tools were essential. As often happens, the

The Nicotinic Acetylcholine Receptor: Current Views and Future Trends,
edited by Francisco J. Barrantes.

history of the AChR has been a tale of fact combined with serendipity. Its modern era began in the early seventies when two unrelated animal species, the electric fish found in the Amazon and the Orinoco rivers, and later the marine electric rays, and the poisonous krait snakes of insular China, were introduced to each other—in the test tube. The former provided an appropriate biological source: electric tissue, the richest source of AChR protein known in nature; the snakes provided α-toxins, 7-8 kDa polypeptides that bind with exquisite selectivity and high to the AChR. A seldom matched pair.

The seventies witnessed advances in the biochemical characterization of the AChR and the "pre-Gigaohm patch-clamp era." Neher and Sakmann[1] developed the patch-clamp technique and first applied it to this particular receptor protein as a test case. The revolution came in the eighties with the development of the full capabilities of the patch-clamp technique, using high-resolution Gigaohm seals[2] combined with molecular genetic approaches leading to the cloning of one, and then all, subunits of the AChR in various laboratories.[3-7] Due credit should be given to the tour de force sequencing of the first 54 amino acids of the *Torpedo californica* α-subunit,[8] which subsequently enabled researchers to develop the oligonucleotide probes leading to identification of clones, initially by hybridization selection, sequencing and, more importantly, the discovery of families and superfamilies of receptors and channels.

When the AChR and several other neurotransmitter receptors entered this "sociological" era by virtue of our understanding of their molecular kindredness, the unexpected conclusion was reached that receptors to pharmacologically "distant" endogenous ligands in fact constitute families and that these, in turn, form the family of evolutionarily related ligand-gated ion channels (LGIC). The latter includes the excitatory nicotinic AChR of skeletal muscle, neurons, and fish electric organs, serotonin-(5-HT_3), γ-amino butyric acid-($GABA_A$) and glycine-activated receptors, extracellular ATP-gated channels, and the sarcoplasmic reticulum ryanodine receptors. The AChR is still the prototype of this superfamily.

The evolutionary relationships between these integral membrane proteins are discussed by Marcelo Ortells in chapter 2. A high degree of sequence homology exists between subunits in a given species, suggesting that they have evolved from a common ancestral gene. Moreover, their primary structure has been remarkably conserved through evolution (80% sequence identity is found for the α subunit between *Torpedo* and man). One extraordinary fact that has come to light as a result of this type of analysis is that LGIC and neuronal-type receptor proteins appeared around 3,000 million years ago, that is, long before the appearance of the nervous system!

The AChR of the Torpedinidae electric organ was the first neurotransmitter receptor to be studied using biochemical techniques. From this standpoint, it is a heterologous transmembrane glycophosphoprotein composed of four different types of subunits assembled as a pentamer of about 300,000 Da, with a stoichiometry of two α1 and one β, γ, and δ in the embryonic or fetal type AChR; in the adult, the γ subunit is replaced by the ε subunit. The pentamer of two α1, and one β, ε, and δ exhibits a briefer channel open time but has a larger conductance. All five subunits are highly homologous, as analyzed in chapter 2 by Ortells, each with four transmembrane domains (M1 to M4). The subunits are arranged like the staves of a barrel around an axis of pseudosymmetry perpendicular to the plane of the membrane, as discussed by Ortells et al in chapter 5 of this volume.

As analyzed by Richard Prince and Steve Sine in chapter 3, the binding of nerve-released acetylcholine (ACh) (two molecules per receptor monomer) to distinct regions of the protein (the α-γ and α-δ interfaces), causes a conformational change in the AChR protein that leads to the opening of its intrinsic cationic channel. The activation of the AChR is terminated within a few milliseconds as ACh diffuses and is hydrolyzed by acetylcholinesterase. Prince and Sine analyze in detail the structural-functional relationships at the AChR ligand-recognition site, the binding domain, with special emphasis on the use of discrete, site-directed (including mutagenesis) modifications of AChR primary structure and their consequences for receptor function.

While agonist recognition occurs at two sites in the extracellular domain of the AChR, the transmembrane domain M2 of each subunit is involved in the ion permeation pathway. Ligand binding triggers conformational changes in the AChR protein that extends to this pathway, the ion pore proper, to cause it to open. In chapter 4 Meyer Jackson makes the transition from ligand recognition to the epiphenomenological consequences of ligand binding by taking a closer look at a crucial question in the field: How does acetylcholine binding affect the functional state of the AChR channel? This issue is addressed by analyzing the thermodynamics of ligand binding and the coupling of binding to AChR channel conformational transitions. Chapter 4 is thus concerned with the application of the to the particular case of the AChR, discussing the energetics of AChR-ligand complex formation, assuming that the energies of specific atomic contacts between ligands and specific amino acid residues within the AChR can be added up to produce specific molecular contacts involved in ligand binding and al transitions between the open and closed channel states.

In chapter 5, Ortells, Georgina Barrantes and myself analyze the current state of the art modeling AChR structure. This theoretical tool is still rather speculative because of the scarcity of hard experimental data on the structure of LGIC. Three-dimensional crystals of the AChR suitable for structural analysis by X-ray diffraction techniques have not been obtained as yet. Elucidation of the high-resolution 3-D structure of the AChR appears not to be just round the corner; even when such a structure does become available, the formidable task of correlating discrete structural conformations of the protein with functional states will barely have commenced.

In the absence of 3-D crystals, structural information on the whole AChR, including its transmembrane region, is currently being gathered by the successful combination of cryoelectron microscopy of two-dimensional lattices of *Torpedo* AChR-rich tubules and image averaging techniques. Two-dimensional crystals of integral membrane proteins suitable for electron diffraction or low-dose cryoelectron microscopy appear to form more readily than 3-D crystals (reviewed in ref. 9). High resolution structural analysis of membrane proteins by single-particle image analysis is also possible, and we applied this technique to the AChR in its native membrane environment almost two decades ago.[10] One of the first demonstrations that the two ligand-binding sites on the AChR are different was obtained with this technique in combination with image averaging techniques.[11]

Nigel Unwin and coworkers (refs. 12, 13 and references therein) have brought the study of ordered 2-D arrays of AChR to its current state by first imaging tubular specimens of *Torpedo marmorata* postsynaptic membranes embedded in amorphous ice (Fig. 1.1A) and then applying averaging techniques to such cryoelectron micrographs. The advances obtained so far permit the observation of well-defined regions of the AChR molecule; rapid freezing of nonliganded and liganded specimens has enabled the observation of differences between the two (Fig. 1.1B).

We can expect significant advances in the obtention of structural data as the extracellular, water-soluble region of the AChR is heterologously expressed in mg amounts in appropriate cellular systems. This will open the way to crystallization of the domain carrying the ligand-recognition site and its study by X-ray diffraction techniques. To date, we have to content ourselves with structural information at 9 Å resolution, which already gives us a good notion of the dimensions and overall shape of the macromolecule (Fig. 1.2).

The AChR appears as a cylindrical body with an overall length of about 120 Å and a diameter of ~65 Å. We can also locate the portions corresponding to the extracellular, membrane-embedded, and cytoplasmic regions of the protein, respectively.

The binding sites of LGIC are much more difficult to study than those for the channel region because detailed structural information for this domain is still lacking. The distinction between the two main structural domains of the AChR, the ligand recognition region and the channel, may have other interesting implications as analyzed in chapters 2 and 5 of this volume. Thus, the ligand-recognition and channel domains may have originally been two different proteins that gradually fused over the lengthy course of evolution.

It is probably the combination of single-channel resolution through the introduction of the patch-clamp technique with the insights provided by genetic engineering (especially site-directed mutagenesis) that has had the clearest impact in the field by disclosing the mechanisms of action of an ever increasing number of ion channels, the AChR included. At the peripheral synapse, Nature ensures a very efficacious mechanism for the activation of the AChR (see chapter 4 in this volume). Firstly, evolution has chosen a small, highly diffusable ligand that binds to a quarter of a million Da protein, the AChR, with a very fast forward rate constant. Secondly, excess ACh is liberated at the neuromuscular junction, thus dictating chemical equilibria also in favor of the biliganded form of the AChR. The rapid binding of the small ligand to its two binding sites on the AChR occurs within a few microseconds (Scheme 1 below):

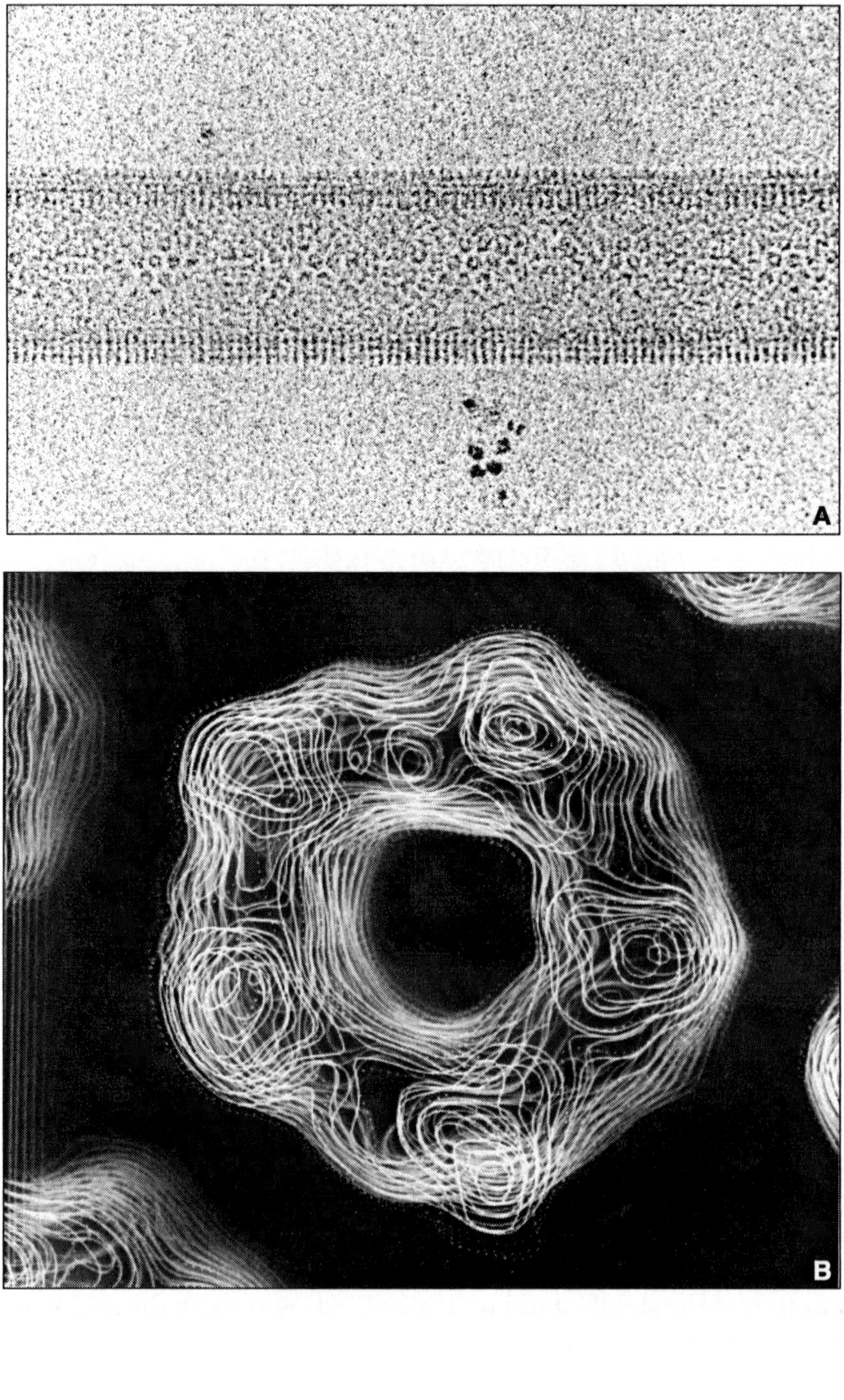

$$2\,A + R \underset{k_{-1}}{\overset{k_1}{\rightleftharpoons}} A + AR \underset{k_{-2}}{\overset{k_2}{\rightleftharpoons}} A_2R \underset{\alpha}{\overset{\beta}{\rightleftharpoons}} A_2R^* \qquad (1)$$

In Scheme 1 (and in its fully developed version in chapter 3 of this volume), two agonist molecules (A) bind to the AChR (R) with association rate constants k_1 and k_2 and dissociation rate constants k_{-1} and k_{-2}. The closed biliganded channel, A_2R, moves to the open configuration (A_2R^*) with an opening rate constant, β, and closes with a closing rate constant, α. The lifetime of the biliganded but closed receptor (A_2R) is brief, and largely determined by the channel closing rate, α, the channel opening rate, β, and the agonist dissociation rate, k_{-2}. The two latter rate constants are of comparable magnitude, thus, the biliganded receptor oscillates several times between the A_2R and A_2R^* states before the agonist dissociates. Alfredo Villaroel discusses in chapter 6 the high degree of sophistication that has already been achieved in the study of ionic conductance through the AChR channel. This has enabled a detailed molecular dissection of the mechanisms of ionic selectivity characteristic of this particular type of channel, the AChR.

As reviewed by Ronald Lukas in chapter 7, only two types of subunits, α (α2-α9) and non-α (β2-β4), have been identified in the CNS and ganglionic neurons in rodents, chicks and humans; thus, diversity of quaternary organization in the nervous tissue seems to be extensive. Unlike the peripheral AChR, the stoichiometry of the CNS receptor has not been definitively established. The α7, α8 and α9 subunits can form homomeric pentamers that produce functional channels when expressed heterologously in promiscuous cell systems

Fig. 1.1. (opposite page) (A) Cryoelectron micrograph of a tubular specimen of *Torpedo marmorata* postsynaptic membrane embedded in amorphous ice. The high order of the specimen can be appreciated at the edges of the tube where the extracellular domains of the AChR face outwards. When this type of specimen is sprayed with acetylcholine either: (i) immediately prior to freezing or (ii) for a prolonged period, differences are observed between the resting conformation and the agonist-induced active (i) and desensitized (ii) conformers, respectively.

(B) The AChR molecule, as viewed from the extracellular, synaptic cleft, in the absence of ACh. Stacked electron density profiles corresponding to sections 2 Å apart obtained by cryoelectron microscopy and image reconstruction techniques from *T. marmorata* samples of AChR-rich membranes. The view corresponds to the "synaptic" mouth of the AChR channel. Illustrations courtesy of Dr. Nigel Unwin. For details see refs. 12-13 and references therein.

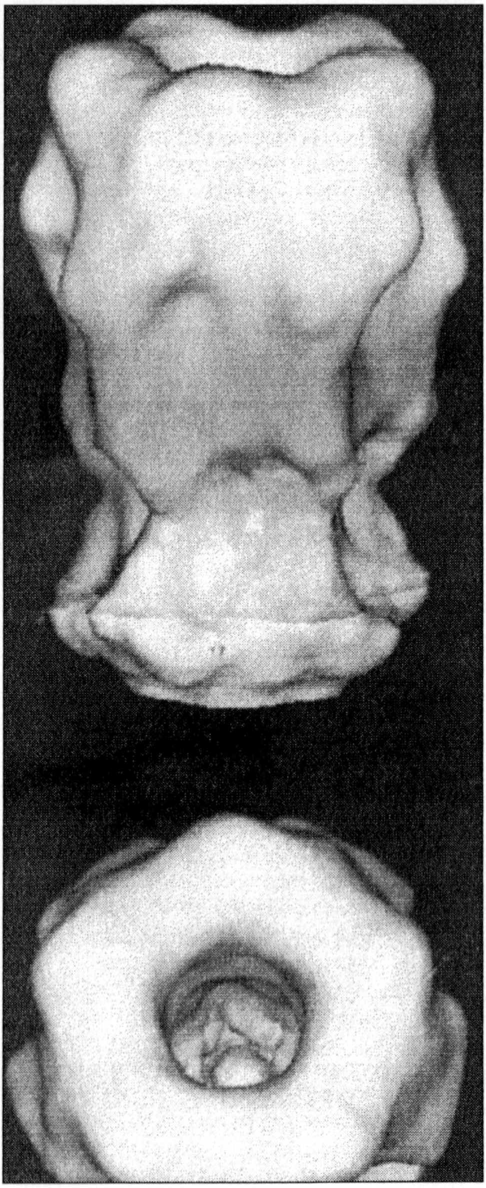

Fig. 1.2. Three-dimensional reconstruction of the AChR molecule as seen from the side (top view) and the synaptic cleft (bottom). The lateral view illustrates the overall cylindrical shape of the 120 Å long molecule, and the distribution of the protein mass exposed to the extracellular milieu, the membrane-embedded domain (banded region) and the much smaller cytoplasmic-exposed domain, respectively. The bottom view enables observation of the vestibule and the mouth of the channel proper, which penetrates the structure in a tunnel-like opening at its center and narrows abruptly after a length of about 60 Å. The three-dimensional model is based on two-dimensional cryoelectron micrographs with a resolution down to 9 Å like those of Fig. 1.1A, reconstructed into a 3-D object by applying image averaging techniques to stacks of 2-D views like those shown in Fig. 1.1B. Illustration courtesy of Dr. Nigel Unwin. For details see ref. 12-13 and references therein. Reprinted with permission from Cell 1993; 72:31-41.

like *Xenopus* oocytes (but with much difficulty and still in an unreproducible fashion in mammalian cells) whereas other α subunits apparently need to be combined with β subunits to form channels. The α9-type is not strictly "neuronal": it occurs in sensory organs, neuroepithelial cells and the tongue and is sensitive not only to nicotinic but also to muscarinic ligands, for which reasons it deserves further investigation. In the same manner that we can today associate the muscle-type AChR with fast synaptic neuromuscular transmission in the peripheral nervous system, it is possible that specific combinations of neuronal nicotinic AChR subunits will be associated with specific brain functions and/or different stages in CNS development and synapse formation. The susceptibility to pharmacological regulation by different ligands may also be associated with different subunit combinations of neuronal AChRs.

The AChR is not exempt from pathological alteration. The same basic principles governing other LGIC apply; the AChR is known to be affected by disease either directly or indirectly. I discuss recent advances in our knowledge of the pathologies of the AChR in chapter 8. Particular emphasis is put on how this recently acquired knowledge may have wider implications that impinge on the structural-functional correlations of these membrane proteins under normal conditions and on pathologies of other ion channels and receptors, and how this information may lead to new diagnostic and therapeutic strategies.

Back to the initial issue. How far are we in the study of the AChR and other ligand-gated channels? The current state of the art in the field is such that we can begin to define the ligand recognition region, the channel proper, or the lipid-facing surfaces of the protein. Though still crude, this level of resolution provides a reasonable framework for rationalizing the structural bases of the resting, closed, and open states of the AChR, the corresponding conformational transitions underlying their interconversion, and the thermodynamics of these processes. Further refinement is needed, with atomic resolution, to fully understand structural-functional relationships, to be able to design appropriate receptor-targeted drugs, and to comprehend the alterations occurring in disease—all in all, to better understand this paradigm molecule of rapid synaptic transmission.

References

1. Neher E, Sakmann B. Single-channel currents recorded from membrane of denervated frog muscle fibres. Nature 1976; 260:779-802.
2. Hamill OP, Marty A, Neher E et al. Improved patch-clamp techniques for high-resolution current recording from cells and cell-free membrane patches. Pflügers Archiv-Eur J Physiol 1981; 391:85-100.
3. Noda M, Takahashi H, Tanabe T et al. Primary structure of alpha-subunit precursor of *Torpedo californica* acetylcholine receptor deduced from cDNA sequence. Nature 1982; 299:793-797.
4. Noda M, Takahashi H, Tanabe T et al. Primary structures of beta- and delta-subunit precursors of *Torpedo californica* acetylcholine receptor deduced from cDNA sequences. Nature 1983; 301:251-255.
5. Noda M, Takahashi H, Tanabe T et al. Structural homology of *Torpedo californica* acetylcholine receptor subunits. Nature 1983; 302:528-532.
6. Ballivet M, Patrick J, Lee J et al. Molecular cloning of cDNA coding for the gamma subunit of *Torpedo* acetylcholine receptor. Proc Natl Acad Sci U S A 1982; 79:4466-4470.
7. Devilliers-Thiery A, Giraudat J, Bentaboulet M et al. Complete messenger RNA coding sequence of the acetylcholine binding α-subunit of *Torpedo marmorata* acetylcholine receptor: A model for the transmembrane organization of the polypeptice chain. Proc Natl Acad Sci U S A 1983; 80:2067-2071.
8. Raftery M, Hunkapiller MW, Strader CD, Hood LE. Acetylcholine receptor: Complex of homologous subunits. Science 1980; 208: 1454-1457.
9. Kühlbrandt W. Two-dimensional crystallization of membrane proteins. Q Rev Biophys 1992; 25:1-49.
10. Zingsheim HP, Neugebauer D-Ch, Barrantes FJ et al. Structural details of membrane-bound acetylcholine receptor from *Torpedo marmorata*. Proc Natl Acad Sci U S A 1980; 77:952-956.
11. Zingsheim HP, Barrantes FJ, Frank J et al. Direct structural localization of two toxin-recognition sites on an acetylcholine receptor protein. Nature 1982; 299:81-84.
12. Unwin N. Nicotinic acetylcholine receptor at 9 Å resolution. J Mol Biol 1993; 229:1101-1124.
13. Unwin N. Acetylcholine receptor channel imaged in the open state. Nature 1995; 373:37-43.

CHAPTER 2

Evolution of the AChR and Other Ligand-Gated Ion Channels

Marcelo O. Ortells

Introduction

The ligand gated ion channel (LGIC) superfamily of receptors is the best known of all the receptor families, predominantly due to the comprehensive characterization of the nicotinic ACh receptor (AChR), which is the paradigm for the whole LGIC superfamily.[1] The AChR and the 5-HT_3 receptors are selective for anions (and hence excitatory) whilst $GABA_A$ and glycine receptors are selective for cations (and are thus inhibitory). Members of the family have a high degree of amino acid sequence homology and share a sequence motif highly characteristic of this group, a 15 residue cys-loop[2] in the N-terminal domain. They are all oligomers, presumably pentamers, and sequence information reveals that for any given member of the family, the subunits are themselves homologous. These subunits have an N-terminal extracellular domain bearing the x site, four putative segments of transmembrane region (M1-M4) and a short extracellular C-terminus. A variable and frequently large cytoplasmic loop lies between transmembrane regions M3 and M4. For further details on LGIC structure, see chapter 5 by Ortells, Barrantes and Barrantes. The ionotropic glutamate receptors were considered for some time as candidates for membership of this superfamily, but this has been ruled out.[3]

The Nicotinic Acetylcholine Receptor: Current Views and Future Trends, edited by Francisco J. Barrantes.

Molecular biology is one of the areas that has contributed much in recent years to the knowledge of LGIC. The sequencing of more than a hundred genes has revealed the presence of closely related receptor families and an unexpected degree of receptor diversity. The receptors are not only widely distributed in phylogenetic terms—they occur from nematodes and insects to vertebrates—but there is also great variability of receptor subunits. This variability is the key element that allows the study of these receptors from an evolutionary point of view.

Evolutionary Trees of the LGIC Superfamily

Interestingly, the latest and more thorough studies on the evolution of LGIC receptors appeared almost simultaneously. So far, only one complete evolutionary tree of the whole LGIC superfamily has been obtained.[4] This work is based on the analysis of nucleotide sequences.[4] Le Novére and Changeux,[5] on the other hand, presented three different evolutionary trees, but these were restricted exclusively to the AChR family. One tree is based on amino acid sequence information, a second on the structure of the AChR genes, and the third is a tentative consensus tree between the two former. Finally, a hypothesis on the evolution of AChR muscle receptors has recently been proposed by Gundelfinger.[6]

DNA Sequence-Based Phylogenetic Tree

Ortells and Lunt[4] employed an alignment of 106 amino acid sequences of LGIC receptors as the starting point for the construction of their evolutionary tree (Fig. 2.1). The initial protein alignment was used as a template for the alignment of the corresponding DNA sequences. The reason for using DNA rather than amino acid sequences for molecular evolutionary analysis is that there are detailed models of the way the former evolves, but none for the latter.[7-9] Detailed models allow, in turn, the construction of evolutionary trees based on more realistic or less arbitrary assumptions. Moreover, because there is no useful evolutionary model for deletions and insertions, only positions shared by all the sequences were considered. Hence, DNA information of only 270 shared codons was used. This eliminated the initial section of each protein, the cytoplasmic loop between the third and fourth transmembrane regions and several other short sections, i.e., regions where there is no obvious homology in the superfamily. Due to the nature of the genetic code, mutations at the third codon position are usually silent (i.e., do not change the amino acid) and consequently have a higher mutation rate. Sequence divergence between some types of LGIC receptors is not small and hence there exists the possibility that superimposed mutations accumulate with time in the third codon position. Since this may lead to inconsistencies in the evolutionary analysis, only the first two codon positions were used.

The method used to build the tree was the maximum likelihood approach,[10,11] using the Felsenstein Phylip 3.5 package.[12] This statistical method looks for the tree that maximizes the probability of the data under the assumption of a given model of DNA evolution. Because there were 106 sequences, it was impossible, for reasons of computational time, to make a thorough search of topologies and calculate for each the likelihood. Instead, a matrix of maximum likelihood distances was calculated and the neighbor-joining[13] and Fitch[14] methods were applied to it. Also, and only to look for alternative topologies, a maximum parsimony algorithm[15] was applied to the original data. A further step was to use only one sequence of each clade (a clade is a group of genes or organisms that share a common ancestor not shared by others) that was consistently composed of the same subunits in the previous analyses. In this case, and because the number of sequences was reduced to 25, a more comprehensive topology search was carried out using the maximum likelihood method proper. After this, the tree was expanded to all the sequences. From all the tree topologies found, the one with the highest likelihood was chosen as the best. Finally, a local rearrangement search for a better tree was applied to this tree. Since the ancestral states of the LGIC superfamily are not known, rooting the tree is rather arbitrary. The choice was to root it at the middle point. This choice appeared to be appropriate as it separated anionic and cationic receptors. Two other topologies were tested against the best found there. The likelihood of best tree, changing the position of the neuronal β2-β4 group by placing it as a sister branch of the neuronal α subunit clade, was calculated. This was done in order to test the alternative hypothesis that neuronal subunits have a common ancestor not shared by muscle subunits. The likelihood of this modified tree was lower than the original best, and hence was no longer considered. The second topology tested was positioning the glycine

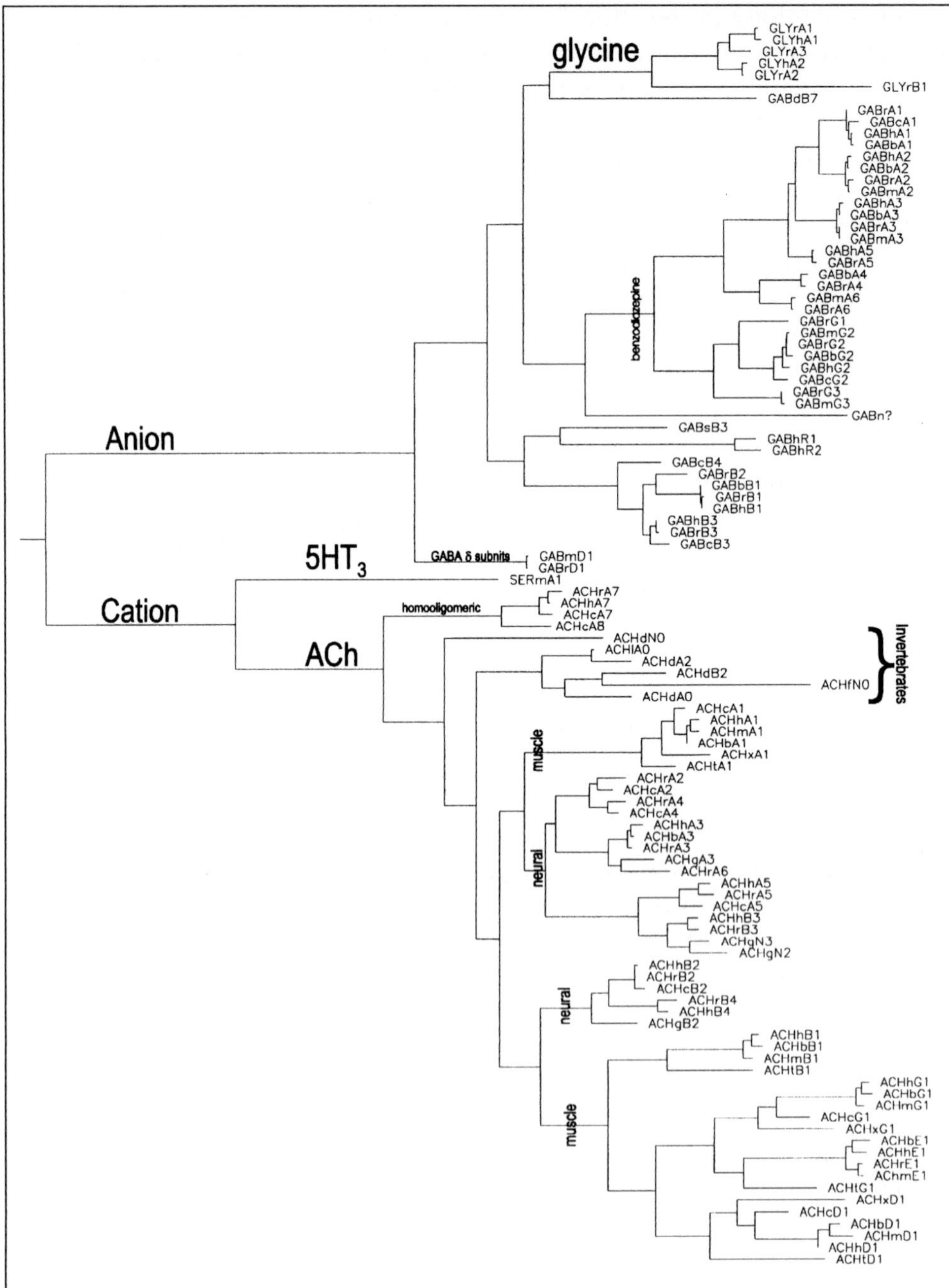
glycine
GLYrA1
GLYhA1
GLYrA3
GLYhA2
GLYrA2
GLYrB1
GABdB7
GABrA1
GABcA1
GABhA1
GABbA1
GABhA2
GABbA2
GABrA2
GABmA2
GABhA3
GABbA3
GABrA3
GABmA3
GABhA5
GABrA5
GABbA4
GABrA4
GABmA6
GABrA6
GABrG1
GABmG2
GABrG2
GABbG2
GABhG2
GABcG2
GABrG3
GABmG3
GABn?
benzodiazepine
Anion
GABsB3
GABhR1
GABhR2
GABcB4
GABrB2
GABbB1
GABrB1
GABhB1
GABhB3
GABrB3
GABcB3
GABA δ subnits
GABmD1
GABrD1
5HT3
SERmA1
Cation
homooligomeric
ACHrA7
ACHhA7
ACHcA7
ACHcA8
ACh
ACHdN0
ACHlA0
ACHdA2
ACHdB2
ACHfN0
ACHdA0
Invertebrates
ACHcA1
ACHhA1
ACHmA1
ACHbA1
ACHxA1
ACHtA1
muscle
ACHrA2
ACHcA2
ACHrA4
ACHcA4
ACHhA3
ACHbA3
ACHrA3
ACHgA3
ACHrA6
neural
ACHhA5
ACHrA5
ACHcA5
ACHhB3
ACHrB3
ACHgN3
ACHgN2
ACHhB2
ACHrB2
ACHcB2
ACHrB4
ACHhB4
ACHgB2
neural
ACHhB1
ACHbB1
ACHmB1
ACHtB1
ACHhG1
ACHbG1
ACHmG1
ACHcG1
ACHxG1
ACHbE1
ACHhE1
ACHrE1
AChmE1
muscle
ACHtG1
ACHxD1
ACHcD1
ACHbD1
ACHmD1
ACHhD1
ACHtD1

Fig. 2.1. (opposite page) Evolutionary tree of the LGIC superfamily.

Symbols used: Six character names of receptors relate to the following nomenclature: RRRsS#, where RRR indicates the type of receptor, s the organism, S the subunit type and # the subunit number (where 0 is undetermined).

Type of receptor, RRR: ACH, Acetylcholine receptor; GAB, GABA receptor; GLY, glycine receptor; SER, $5HT_3$ receptor.

Organism, s: b, bovine; c, chicken; d, *Drosophila*; f, filaria; g, goldfish; h, human; l, locust; m, mouse; n, nematode; r, rat; s, snail; t, *Torpedo*; x, *Xenopus*. Subunit type, S: A, alpha; B, beta; G, gamma; D, delta; E, epsilon; R, rho; N, non-α; ?, undetermined.

Database accession numbers for the sequences. Code: accession numbers beginning with K, L, M, and X are from EMBL database, those beginning with A, B, JH, JN, JQ, and S are from PIR; and those beginning with J are from GeneBank. Reprinted with permission from Ortells MO and Lunt GG, Trends Neurosci 1995; 18:121-127.

ACHbA1:X02509;	ACHbA3:X57032;	ACHbB1:X00962;	ACHbD1:X02473;
ACHbE1:X02597;	ACHbG1:M28307;	ACHcA1:X12434;	ACHcA2:X07340-4;
ACHcA4:X07348-52;	ACHcA5:J05643;	ACHcA7:X52295;	ACHcA8:JH0173;
ACHcB2:X07353-7;	ACHcD1:K02903;	ACHcG1:K02904;	ACHdA0:X07194;
ACHdA2:X52274;	ACHdB2:S12899;	ACHdN0:S03012;	ACHfN0:L12543;
ACHgA3:X54051;	ACHgB2:X54052;	ACHgN2:X14786;	ACHgN3:M29529;
ACHhA1:X17104;	ACHhA3:A37040;	ACHhA5:M83712;	ACHhA7:L25827;
ACHhB1:X14830;	ACHhB2:X53179;	ACHhB3:X67513;	ACHhB4:X68275;
ACHhD1:X55019,	X53091-516;	ACHhE1:X66403;	ACHhG1:X01715-21;
ACHlA0:S12359;	ACHmA1:X03986;	ACHmB1:M14537;	ACHmD1:X13959;
ACHmE1 X55718;	ACHmG1:M30514;	ACHrA2:M20292-7;	ACHrA3:X03440;
ACHrA4:M15681;	ACHrA5:A35721;	ACHrA6:L08227;	ACHrA7:M85273;
ACHrB2:JH0174;	ACHrB3:A33523;	ACHrB4:B35721;	ACHrE1:X13252;
ACHtA1:X13252;	ACHtB1:A03171;	ACHtD1:A03177;	ACHtG1:A03173;
ACHxA1:X17244;	ACHxD1:X07069;	ACHxG1:X07068;	GABbA1:X05717;
GABbA2:X12361;	GABbA3:X12362;	GABbA4:X61456;	GABbB1:X05718;
GABbG2:M55563;	GABcA1:X54244;	GABcB3:X54243;	GABcB4:X56647;
GABcG2:X54944;	GABdB7:M69057;	GABhA1:X13584;	GABhA2:S62907;
GABhA3:S62908;	GABhA5:L08485;	GABhB1:X14767;	GABhB3:M82919;
GABhG2:X15376;	GABhR1:M62400;	GABhR2:M86868;	GABmA2:M86567;
GABmA3:M86568;	GABmA6:X51986;	GABmD1:M60587;	GABmG2:M57522;
GABmG3:X59300;	GABn?:X73584;	GABrA1:S03889;	GABrA2:JH0370;
GABrA3:A34130;	GABrA4:S17551;	GABrA5:B34130;	GABrA6:L08495;
GABrB1:X15466;	GABrB2:X15467;	GABrB3:X15468;	GABrD1:M35162;
GABrG1:S12056;	GABrG2:JQ0077;	GABrG3:M81142;	GABsB3:X58638;
GLYhA1:S12382;	GLYhA2:S12381;	GLYrA1:A27141;	GLYrA2:JN0112;
GLYrA3:M55250;	GLYrB1:JH0165;	SERmA1:M74425	

receptors as a sister branch of GABA receptors, and as before, the tree obtained had a lower likelihood than the original best. The same analyses were made excluding the transmembrane regions and the same results were obtained. Transmembrane regions are more restricted in their amino acid compositions because hydrophobic residues are better suited for the lipid environment, and this could be the source of convergence (i.e., the same amino acid in a particular position but originating from a different mutation) that can produce wrong tree topologies. However, it is also true that in the extracellular domain, there are positions that are functionally or structurally restricted in their amino acid composition, and thus probably the extracellular and transmembrane domains are equally informative.

The tree in Figure 2.2 has the same topology as the one in Figure 2.1, but branch lengths were recalculated assuming a molecular clock. The use and fundamentals of the molecular clock hypothesis have been and are still controversial. It was discovered that certain proteins had a constancy in the rate of amino acid substitutions among several mammalian lineages. Consequently, it was suggested[16] that the rate of molecular evolution of a given protein is almost constant over time, hence the term "molecular clock." The importance of this discovery is that proteins could be used to date divergence times of species (or genes) in a way similar to that of isotope employment. A molecular clock is not needed for reconstructing phylogenetic trees, and as long as backward and parallel mutations are uncommon, these can be now estimated quite effortlessly. To estimate divergence times using molecular information, a known time scale is needed to convert relative molecular distances to real time. Fossil information is employed for this purpose, although it may not be precise.

Fig. 2.2. (opposite page) Evolutionary tree of the LGIC superfamily assuming a molecular clock. The topology is the same as in Fig. 2.1, but the distance from the ancestor to any tip (present day receptor) is the same, or in other words, all subunits have evolved at the same rate. For symbol explanation see Fig. 2.1. Calibration from relative to absolute time scaling was based on the fossil record for the average time of divergence of the lineages leading to: mammals and birds (approximately 300 Myr ago), mammals-birds and amphibians (approximately 350 Myr ago), and fish and the remaining vertebrates (approximately 430 Myr ago, Løvtrup[37]). The error is ±550 Myr. Reprinted with permission from Ortells MO and Lunt GG, Trends Neurosci 1995; 18:121-127.

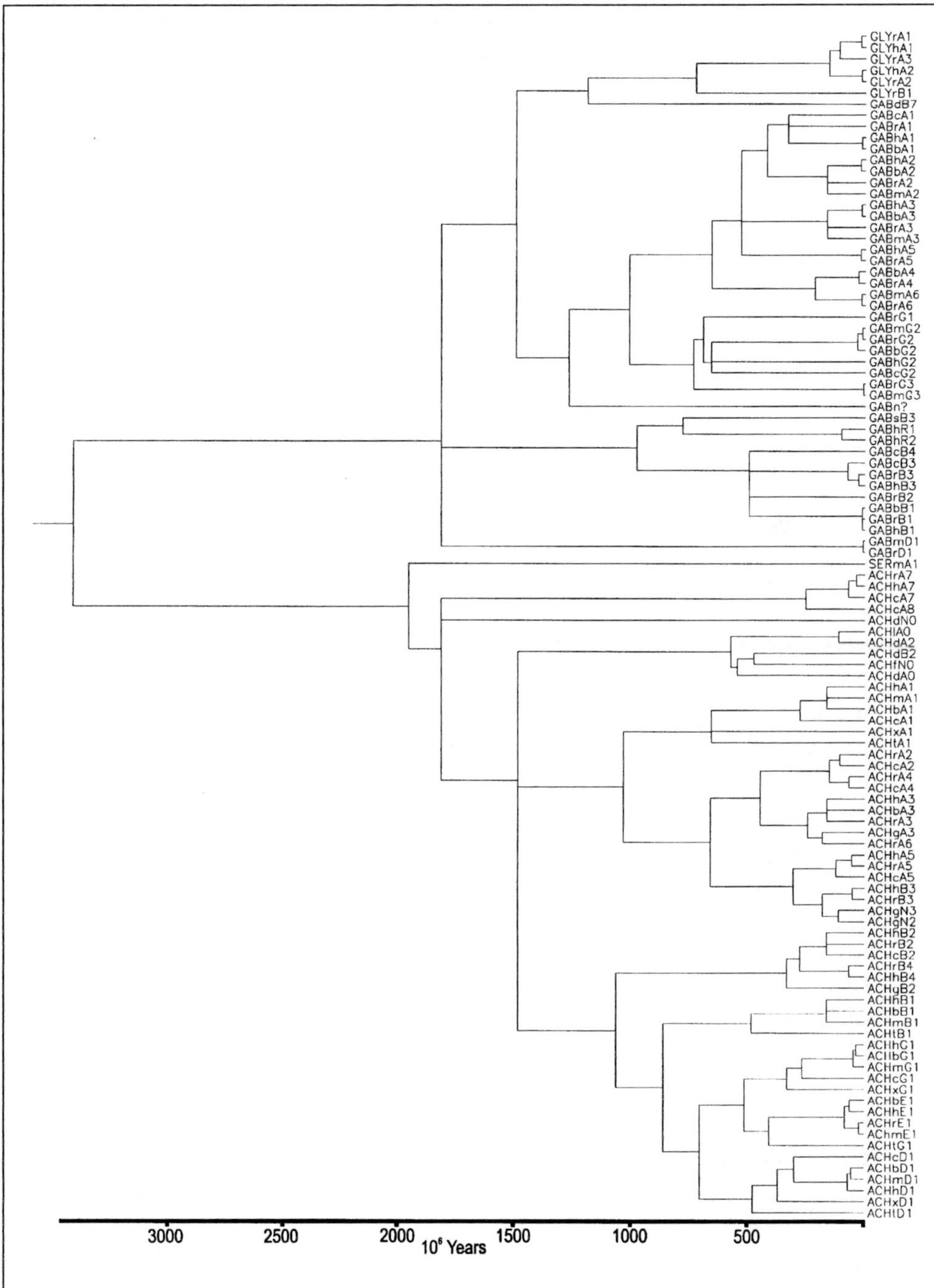
GLYrA1
GLYhA1
GLYrA3
GLYhA2
GLYrA2
GLYrB1
GABdB7
GABcA1
GABrA1
GABhA1
GABbA1
GABhA2
GABbA2
GABrA2
GABmA2
GABhA3
GABbA3
GABrA3
GABmA3
GABhA5
GABrA5
GABbA4
GABrA4
GABmA6
GABrA6
GABrG1
GABmG2
GABrG2
GABbG2
GABhG2
GABcG2
GABrG3
GABmG3
GABn?
GABsB3
GABhR1
GABhR2
GABcB4
GABcB3
GABrB3
GABhB3
GABrB2
GABbB1
GABrB1
GABhB1
GABmD1
GABrD1
SERmA1
ACHrA7
ACHhA7
ACHcA7
ACHcA8
ACHdN0
ACHlA0
ACHdA2
ACHdB2
ACHfN0
ACHdA0
ACHhA1
ACHmA1
ACHbA1
ACHcA1
ACHxA1
ACHtA1
ACHrA2
ACHcA2
ACHrA4
ACHcA4
ACHhA3
ACHbA3
ACHrA3
ACHgA3
ACHrA6
ACHhA5
ACHrA5
ACHcA5
ACHhB3
ACHrB3
ACHgN3
ACHgN2
ACHhB2
ACHrB2
ACHcB2
ACHrB4
ACHhB4
ACHgB2
ACHhB1
ACHbB1
ACHmB1
ACHtB1
ACHhG1
ACHbG1
ACHmG1
ACHcG1
ACHxG1
ACHbE1
ACHhE1
ACHrE1
AChmE1
ACHtG1
ACHcD1
ACHbD1
ACHmD1
ACHhD1
ACHxD1
ACHtD1
3000
2500
2000
1500
1000
500
10⁶ Years

AChR Amino Acid Sequence-Based Phylogenetic Tree

Le Novére and Changeux[5] analyzed the amino acid sequences of the subunits belonging to the nicotinic family of receptors, that is, a subtree of the whole LGIC superfamily analyzed by Ortells and Lunt.[4,17] They used two procedures, maximum parsimony[15] and the phenetic neighbor-joining method,[13] for constructing evolutionary trees. For the former, an alignment with 48 sequences with 357 informative sites was used, and a distance matrix based on the Dayhoff[18] PAM matrix was used for the latter. Mouse serotonin and rat α3 glycine subunits were used as outgroups.

Structure-Based AChR Muscle Subunit Tree Topology

Le Novére and Changeux[5] also estimated an evolutionary tree based on the information of the structure of the subunit genes and using the maximum parsimony method. Because this tree was incompatible in its topology with the one based on amino acid sequences, they constructed a summary tree (shown here in Fig. 2.3) where they tentatively accommodated the phylogenetic information available (see below). A third alternative topology for the evolution of the AChR muscle subunits, proposed by Gundelfinger[6] and schematically presented in Figure 2.4, will be discussed later when comparing all the topologies.

Origin of the LGIC Superfamily

By rooting the DNA based tree in the middle of its length, the ancestor of the LGIC is placed between cationic and anionic receptors. Assuming a molecular clock, the date estimated for this ancestor is at least 2500 Myr ago.[4] The first impression might well be that this is a surprisingly remote origin, probably before the first eukaryotes.[19] However, this seems not to be an isolated case. In spite of the lack of sequence similarity, G-coupled protein receptors, another major group of cell surface signaling proteins, are known to have a tertiary structure similar to bacteriorhodopsin[20] and are probably homologous to what is clearly a prokaryotic protein. Such considerations suggest that these very important surface signaling molecules associated with present day nervous systems were readily available well before this novel signaling function made its appearance during evolution. An earlier study[21] suggested that LGIC-like proteins may be widely represented in a variety of organisms, and the ancestral

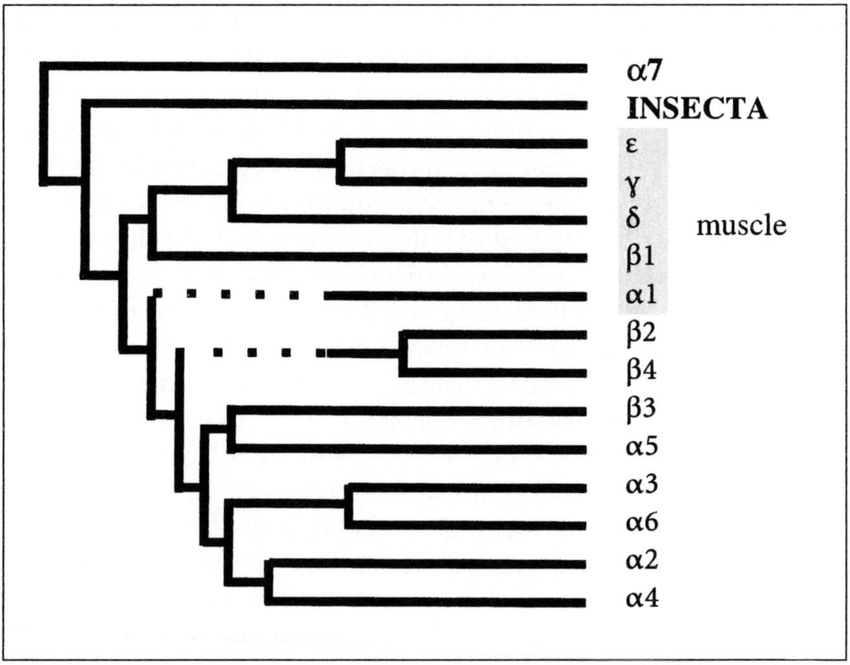

Fig. 2.3. Consensus evolutionary tree for the AChR family. Based on amino acid sequence data and gene structure. Redrawn from Le Novère N et al, J Mol Evol 1995; 40:155-172.

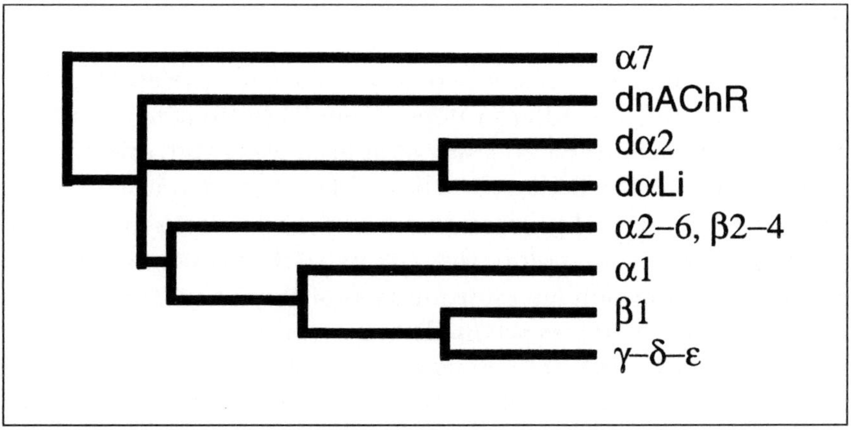

Fig. 2.4. Most parsimonious tree based on gene structure information.

role of primitive LGIC receptors was discussed in the context of osmotic regulation and nutrient seeking, both of which may involve transmembrane ion fluxes and ligand recognition. Nutrient seeking may relate to the present function of the LGIC since their extracellular region is a highly complex molecular recognition system that would not be needed for osmotic regulation, a function that might more probably relate to that of voltage-dependent ion channels.

The phenomenon of desensitization[22] at the molecular level (in contrast to the rather confusing and diverse concept of cellular desensitization) can also be considered in the context of these ancient putative origins of the LGIC.[4] Desensitization may have provided a mechanism to prevent a long term opening of the channel that could not only be harmful, but futile to the primitive cell. If acetylcholine reaches high enough concentrations in the synapse to provoke receptor desensitization, this phenomenon might be a normal and useful part of synaptic transmission. If not, there would have been no selectivity pressure to keep this feature and this would indicate that desensitization is intrinsically inherent in the very basics of the structure of these receptors, and probably strongly selected for prior to the acquisition by the LGIC of their present function in the nervous system. Dudel and Franke[23] summarized our knowledge on the roles of desensitization in several synapses. They stated that in the vertebrate neuromuscular junction, receptor desensitization does not play any direct role in shaping the synaptic current. Hence, it seems plausible that this characteristic originated in the ancestral proteins of the present-day LGIC.[17]

So far, there is no evidence of any evolutionary relationship between LGIC and any other protein. Given the estimated divergence time from the common ancestor of at least 2500 Myr, this may not be surprising. Nevertheless, Ortells and Lunt[4] tried to use the information from the evolutionary tree they found to get some clues to the origins of these receptors; they estimated the hypothetical state(s) of the LGIC common ancestor for each of the 810 nucleotide positions analyzed. This was accomplished by using the topology of the tree and a maximum parsimony estimation of the ancestral states. This ancestral sequence, which obviously has several redundancies, was submitted to the Daresbury Laboratory DNA database for a search of similarities. Interestingly, within the 300 best scores, only 5-HT_3 and GABA δ-1 subunits were selected from the LGIC superfamily, suggesting an "early" nature for these two receptors. In this analysis only those sequences that had optimized scores higher than the receptor sequence that scored worst were considered. All of them, except one, coded for highly repetitive proteins or repetitive DNA (which by chance alone can have approximately 25% identity with the probe). The exception was a nonrepetitive protein whose reading frame matched that of the probe. This protein is a beta-ketoacyl synthase from *Streptomyces glaucescens* (accession code x15312). However, the amino acid sequence alignment with the $5HT_3$ and GABA δ1 subunits is not significant[24] (percentage identity between 11 and 12%), although a very conserved region in LGIC, the 15 amino acid cys-loop, seems to be present. Nevertheless, the two proteins β-ketoacyl synthase and LGIC are not evolutionary related.

In another work Ortells[38] attempted to thread the extracellular receptor sequence using the methods and software available at the PredictProtein server at EMBL, Heidelberg. The highest z-score obtained was 2.36, and only with z-scores higher than 3.5 is the first hit correct in at best in 60% of cases. Hence, it seems that for the method used, there is no known protein structure to which the receptor sequence can be folded. Tertiary structures can be maintained between different proteins, even where there is no obvious DNA or amino acid sequence similarity. Therefore, and as before, it seems that LGIC are not related to any known protein sequence or structure available in the databases.

Although it can be argued that the failure to find a LGIC related protein might be due to the incompleteness of these databases, it is more probable that either the sequence signal is no longer recognizable or that LGIC are made of two or more unrelated proteins joined by exon shuffling. For example, Unwin[25] noted a probable analogy between the structure of the transmembrane region of LGIC and two related pentameric bacterial toxins, heat-labile enterotoxin[26] and verotoxin.[27] These two toxins have an obvious similarity in their tertiary structure but not at the sequence level.[27] Hence, it may be quite feasible that the transmembrane region of LGIC is derived from a toxin-like related protein,[28,29] whilst the extracellular domain has a different origin.

One last point to emphasize is the lack of rationale in assigning any degree of primitiveness to receptors belonging to 'primitive' organisms. It can be said that an organism is primitive in terms of the moment in the evolutionary time scale at which it appeared. Bacteria are primitive organisms in the sense that all the characteristics that make and constrain them genetically and thermodynamically to remain as bacteria appeared very early. By the same token, a LGIC receptor is a very primitive protein; it has been a receptor for perhaps 2500 Myr. However, a vertebrate AChR can be considered more 'primitive' than an insect AChR if it shares more attributes with the original protein; that can depend on both selection pressures and on chance (genetic drift). Present day receptors from 'primitive organisms' have evolved over the same period of time as those from mammalian brain. As we saw, $5HT_3$ receptors or GABA δ subunits are probably the most primitive of all receptors, as they are more similar to the hypothetical ancestor, regardless of whether they belong to vertebrates or insects.

Evolution of Cationic Receptors

Within the cationic receptors, the $5\text{-}HT_3$ and AChRs diverged very early. As seen from the tree in Figure 2.1, the divergence of $5\text{-}HT_3$ receptors with respect to the cationic ancestor is the lowest, indicating that they could be the most similar to the primitive cationic receptor.[4]

According to Figures 2.1 and 2.2, among AChR subunits the most primitive (considering the length of their branch) and the ones that split early from the remainder are the α7 and α8 subunits, which are capable of forming homooligomeric receptors in *Xenopus* oocytes.[39] After the α7 split, there followed some α or α-like subunits cloned from invertebrates (this does not mean that they cannot be present in vertebrates). It is known that at least one of these subunits (ACHIA0) can also form homooligomeric receptors[40] in the *Xenopus* expression system. Given the primitiveness of the α7 subunit, its closeness to $5\text{-}HT_3$ receptors, which are also homooligomeric, and the probable fact that there were no other subunits when the α7-type AChR appeared, it is highly likely that these receptors may be truly homooligomeric in nature.

Nicotinic AChR α subunits are defined by the possession of a pair of adjacent cysteines in the N-terminal domain that are believed to be involved in ligand binding. Indeed, ligand binding in AChRs has been generally thought to be mainly associated with the α subunits[30] (but see chapter 3). As seen from the trees of Figures 2.1 and

2.2, non-α or structural subunits (lacking the pair of cysteines) have derived from α subunits at three independent times. First, there are non-α subunits within the 'invertebrate' group of receptors. Second, there is a main division between α and non-α (see below). Third, within the last α group, and as a sister branch of the α5, the β3 subunits (including the goldfish N3 and No) appeared.

Probably at about the same time, α and β subunits (with the exception of the β3 that have an independent origin) split into their neural and muscle subtypes (Figs. 2.1 and 2.2). This could have been caused by the separation of both tissues. The minimum date for this event is between 800 and 1400 Myr ago. In both cases (α and non-α), the neural subunits seem to be more similar to their respective ancestors than their muscle counterparts (Fig. 2.2). This means that until the gene duplication and subsequent specialization of the muscle subunits, neural and muscle receptors were probably the same.

Within the non-α group, the neural type had only a minor bifurcation (β2 and β4), although the goldfish β2 receptor seems to be something different. However, the muscle types split several times, giving rise to very different subunits. The β subunits separated early and are also the most similar to the ancestor. The other branch split into the δ on the one hand and the γ and ε on the other. What is called a γ subunit in *Torpedo* is in fact more related to the ε of other vertebrates. In the case of mammals, the ε subunit replaces the γ in the adult.

Nevertheless, there are alternative views for the evolution of AChRs. Le Novére and Changeux[5] obtained for their amino acid sequence analysis the same results as Ortells and Lunt.[4] However, they also used the gene structures of the AChRs as a source of evolutionary information. The tree obtained from these data using maximum parsimony was not fully congruent with those achieved with sequence (either DNA or amino acid) information. In this tree, vertebrate subunits were not split, as above, between α and non-α, but were divided between muscle and neuronal subunits. In order to deal with this inconsistency, they constructed a hybrid tree (Fig. 2.3) where they placed the α1 (muscle) subunit within the neuronal group, the latter now including also the β2-β4 subunits formerly joined to the muscle non-α subunits (as in ref. 4).

A third point of view was proposed by Gundelfinger.[6] He suggested that the muscle receptor subunits evolved monophyletically from a common ancestor related to the α1 subunit. This view is exactly the one suggested by the tree obtained with the information of the gene structures.[5] Two other sources of information were included to reinforce his hypothesis: 1) the position of two potential N-glycosylation sites, one of which is well conserved in the α7-α8 group, the invertebrate receptor subunits and nearly all the neuronal subunits, but absent in the muscle subunits; 2) according to Gundelfinger,[6] the mutation rate of vertebrate muscle β1, γ and δ subunits is more than twice as that of the muscle α1; therefore, the divergence time between muscle α and non-α subunits might be much shorter than the indicated by sequence divergence. Moreover, citing Le Novére and Changeux,[5] he remarked that the sequence conservation between β2-β4 vertebrate neuronal subunits and muscle non-α subunits is not significant and could represent a convergence driven by functional constraints. However, these arguments are not strong enough to fully support his hypothesis. All the muscle subunits share two introns (one between M1 and M2, and the other after M3) not shared either with the insect (neural) subunits or the vertebrate neural subunits. Two alternative topologies can be congruent with these facts. The first, as proposed by Gundelfinger,[6] has the muscle subunits in one group and the neural subunits in another. However, the average of pairwise comparisons between muscle and neural α subunit gives a 45% percentage identity, higher even than the average between α neural (including the β3 subgroup) and non-α neural (43%), whereas the average percentage identity between muscle α and muscle non-α is 33%. Also, the comparison between the β2-β4 group to the muscle non-α gives an average percentage identity of 38%, that is, higher than the muscle α/non-α comparison. The other alternative is to position the β2-β4 group as a sister branch of the neural α group as done by Le Novére and Changeux,[5] but this topology also does not account for the above facts.

Gundelfinger,[6] when quoting Le Novére and Changeux,[5] pointed out that the node connecting the β2-β4 group to the muscular non-α subunits is not significant, but did not take into account that this estimation is based on the less informative protein sequence data, for which there is no model of evolution. Ortells'and Lunt's[4,17] data, based on DNA sequence, clearly indicates, on the contrary, that both the nodes connecting the β2-β4 and muscle non-α subunits, and the muscle α1 to the neuronal α subunits are significant at the 0.01 level. Furthermore, the advocacy of evolutionary convergence to account for the sequence similarity between the β2-β4 and muscular non-α subunits cannot explain why all the other non-α subunits did not also converge to the same amino acid sequence. Hence, each of these hypotheses has some inconsistencies; most probably the origin of the inconsistencies lies in the type of data used in each case.

The information from the structures of the genes (i.e., number and position of introns) poses several problems of interpretation in the context of understanding the evolution of these receptors. There is no clear understanding of the way introns evolve (in contrast to DNA sequence evolution). For example, the introns between M1 and M2 in the muscle α and β subunits are in a position slightly different from those of the δ, γ, and ε subunits, which might mean either that the genes are not homologous, or that introns can change positions. If the latter is true, then it is also possible, for example, that the intron between M1 and M2 of the neuronal ARD subunit of *Drosophila*[6] could be homologous to the muscle intron, in which case intron-position data gives no useful information in our case.

Gundelfinger[6] reinforced his hypothesis with the observation of the presence of two conserved glycosylation sites. One, close to the cys-loop, is also present in the 5-HT_3 and $GABA_A$ receptors and hence is not informative. The other, absent in all the muscle AChR subunits, is also absent in other neuronal AChR subunits. The amino acid and DNA sequence of this region gives no indication of a closer relationship between the muscle α and non-α subunits, with respect to neuronal subunits that also lack this glycosylation site. Moreover, two residues, D and R, present in some muscle α subunits in the first and third positions, are also present in the 5-HT_3 receptor, possibly indicating the poor informative value of this site.

Evolution of Anionic Receptors

Within anionic receptors, the most significant fact is that glycine receptors are derived from $GABA_A$ receptors. At first this was surprising given the fact that GABA is a rather complex molecule as compared with the simple and common amino acid glycine. It is often assumed that during evolution readily available molecules such as glycine were the first to be used as transmitters.[31] However, it is also possible that in the primitive environment, with less complex topologies or compartmentalization, such a very common molecule carried no unique information. As mentioned before, probably the most primitive of the anionic receptors is the GABA δ1. An important feature of GABA receptors is the presence of several modulatory sites, which in response to a variety of ligands (e.g., benzodiazepines), can allosterically interact with the GABA binding site.[32] It is generally agreed that the recognition of GABA is primarily associated with the β subunits and these are seen in both invertebrates and mammals. It is also observed that within the $GABA_A$ receptor family those subunits that are considered to be primarily concerned with the benzodiazepine responses, α and γ, belong to the sister group of the β-class of subunit but they are certainly not temporally linked to the appearance of the vertebrates. The presence of non-β subunits in the invertebrates is not yet well established.

Why So Many Receptors?

Within the main nicotinic α-subunit group, only the neuronal AChRs have proliferated into many subtypes, and if we assume a molecular clock this multiplication took place very recently. It seems that the non-α subunit proliferation in the muscle may have been a means to "fine tune" a single role based on one receptor, whilst in the brain it was used to expand beyond one role, thus generating different receptors. In the case of the anionic receptors we see a similar proliferation of subtypes, with, in the case of mammalian brain $GABA_A$ receptors, six αs, four βs, three γs and two ρs. Was this type of evolution positively selected or was it the result of gene drift?

I tried to test (for both anionic and cationic receptors) the possibility of selection favoring these diversifications. This can be accomplished by comparing the rates of substitutions in synonymous (K_s) and nonsynonymous (K_a) nucleotide positions[33,34] in subunits that have a recent common ancestor. Usually it is found that K_s is greater than K_a.[35] If the rate of nonsynonymous substitutions is higher than the synonymous, positive natural selection can be inferred. In muscle we might expect to find selection pressure in regions other than the binding site because mainly α subunits bear them and these have never changed in this tissue. In other words, if selection had occurred in the binding site region, we would expect more proliferation of different α subunits (as a consequence of selection) because of their central role in ligand binding. This does not hold for neuronal subunits where both α and β subunits have proliferated. For these reasons, two specific regions were tested using Li's[36] method, the extracellular domain (where the binding site is located) and M2 (the transmembrane region forming the ion channel).

In all the cases, K_s values were significantly higher (data not shown), giving no indication of selection. This does not rule out selection entirely because it is possible that the time elapsed since the gene duplications may have been enough to override any signal of selection. However, there is an interesting case. The comparison between the M2 regions of *Torpedo* β1 and bovine, mouse and human δ1 subunits (subunits belonging to sister groups) gave an indication (although not statistically significant) of lower K_s, or at least equal K_s and K_a values. This is also surprising given the amount of divergence between the subunit subtypes. A possible explanation is that positive selection has occurred in *Torpedo* as a way of modifying the normal muscle AChR towards its new "electrical" function. Actually, we do not know how different the AChR in *Torpedo* muscle is from that in the electric tissue as no data are found in the literature. Analyzing the M2 region for any particularity of the *Torpedo* sequence, I found a valine residue at position 13 of the alignment of Table 2.1, that is exclusive to this genus (none of the remaining M2 residues are unique to *Torpedo*). This valine is not found in any other LGIC M2 sequence suggesting that it might play an as yet undefined important role in the channel properties that relates particularly to the "electrical" function of the *Torpedo* electroplax. This position is occupied with Gln or Leu in all other nicotinic subunits. Serotonin receptor has a Tyr, and anion channels have either Met or Leu in this position. It would be worthy of study a mutation of this valine to either Leu or Gln. In this context it is a pity that there are no full length sequences available for the AChR from the electric organ of a quite different fish, that of the fresh water eel *Electrophorus electricus*.

The lack of evidence for positive selection in neuronal AChR subunits may also relate to the fact that I tested for the whole extracellular region. Selection may have acted only on some important specific regions such as the binding site, but we do not know exactly where to look in the sequence, mainly because several linearly unconnected pieces of sequences may be highly associated in the three-dimensional structure and constitute the overall "binding site," as analyzed in more detail in chapter 3 by Prince and Sine.

Table 2.1. Sequence alignment of the muscle δ1 M2 region

	1									10										20		
Human δ1	E	K	M	G	L	S	I	F	A	L	L	T	**L**	T	V	F	L	L	L	L	A	D
Bovine δ1	E	K	M	G	L	S	I	F	A	L	L	T	**L**	T	V	F	L	L	L	L	A	D
Mouse δ1	E	K	M	G	L	S	I	F	A	L	L	T	**L**	T	V	F	L	L	L	L	A	D
Torpedo δ1	E	K	M	S	L	S	I	S	A	L	L	A	**V**	T	V	F	L	L	L	L	A	D

References

1. Unwin N. Neurotransmitter Action: Opening of ligand-gated ion channels. Cell 1993; 72:31-41.
2. Cockcroft VB, Osguthorpe DJ, Barnard EA et al. Modeling of agonist binding to the ligand-Gated ion channel superfamily of receptors. PROTEINS: Struc Func Genet 1990; 8:386-397.
3. Sutcliffe MJ, Wo G, Oswald RE. Three-dimensional models of non-NMDA glutamate receptors. Biophysical J 1996; 70:1575-1589.
4. Ortells MO, Lunt GG. Evolutionary history of the ligand-gated ion-channel superfamily of receptors. Trends Neurosci 1995; 18:121-127.
5. Le Novère N, Changeux J-P. Molecular evolution of the nicotinic acetylcholine receptor: An example of multigene family in excitable cells. J Mol Evol 1995; 40:155-172.
6. Gundelfinger ED. Evolution and desensitization of LGIC receptors. Trends Neurosci 1995; 18:297.
7. Jukes TH, Cantor CR. Evolution of protein molecules. In: Munro, HN, ed. Mammalian Protein Metabolism. New York: Academic Press, 1969:21-132.
8. Jin L, Nei M. Limitations of the evolutionary parsimony method of phylogenetic analysis. Mol Biol Evol 1990; 7:82-102.
9. Kimura MJ. A simple method for estimating evolutionary rate of base substitution through comparative studies of nucleotide sequences. Mol Evol 1980; 16:111-120.
10. Felsenstein J. Evolutionary trees from DNA sequences: A maximum likelihood approach. J Mol Evol 1981; 17:368-376.
11. Saitou N. Maximum likelihood methods. Methods Enzymol 1990; 183:584-598.
12. Felsenstein, J. PHYLIP (Phylogeny Inference Package) version 3.5c. Distributed by the author. Department of Genetics, University of Washington, Seattle; 1993.
13. Saitou N, Nei M. The neighbor-joining method for reconstructing phylogenetic trees. Mol Biol Evol 1987; 4:406-425.

14. Fitch WM, Margoliash E. Construction of phylogenetic trees. A method based on mutation distances as estimated from cytochrome c sequences is of general applicability. Science 1967; 155:279-284.
15. Fitch WM. Toward defining the course of evolution: Minimum change for specific tree topology. Syst Zool 1971; 20:406-416.
16. Zuckerkandl E, Pauling L. Evolutionary divergence and convergence in proteins. In: Bryson V, Vojel HJ eds. Evolving Genes and Proteins. New York: Academic Press, 1965:97-166.
17. Ortells MO, Lunt GG. Evolution and desensitization of LGIC receptors. Trends Neurosci 1995; 18:297-299.
18. Dayhoff, MO. Atlas of protein sequence and structure, vol. 5, supplement 3, 1978. National Biomedical Research Foundation, Washington DC.
19. Schopf JW. Earth's Earliest Biosphere: Its Origin and Evolution. Guildford N.J.:Princeton University Press, 1983.
20. Henderson R, Baldwin JM, Ceska TA et al. Model for the structure of bacteriorhodopsin based on high-resolution electron cryo-microscopy. J Mol Biol 1990; 213:899-929.
21. Cockcroft VB, Osguthorpe DJ, Barnard Eaet al. Ligand-gated ion channels: Homology and diversity. Molecular Neurobiology 1990; 4:129-169.
22. Changeux JP, Linás RR, Purves D, Bloom FE, eds. Fidia Research Foundation Neuroscience Award Lectures Volume 4. Raven Press, Ltd, 1990:21-168.
23. Dudel J, Franke C. Evolution and desensitization of LGIC receptors. Trends Neurosci 1995; 18:297-298.
24. Sander C, Schneider R. Database of homology-derived protein structures and the structural meaning of sequence alignment. PROTEINS: Struc Func Genet 1991; 9:56-68.
25. Unwin N. Nicotinic acetylcholine receptor at 9 Å resolution. J Mol Biol 1993; 229:1101-1124.
26. Sixma TK, Pronk SE, Kalk KH et al. Crystal structure of a cholera toxin-related heat-labile enterotoxin from *E. coli.* Nature 1991; 351:371-377.
27. Stein PE, Boodhoo A, Tyrrell GJ et al. Crystal structure of the cell-binding B oligomer of verotoxin-1 from *E. coli.* Nature 1992; 355:748-750.
28. Ortells MO, Lunt GG. A mixed helix-beta sheet model of the transmembrane region of the nicotinic acetylcholine receptor. Prot Engng 1996; 9:51-59.
29. Ortells MO, Barrantes GE, Wood et al. Molecular modelling of the nicotinic acetylcholine receptor transmembrane region in the open state. Prot Engng 1997; (in press).
30. Karlin A. Structure of nicotinic acetylcholine receptors. Curr Opin Neur 1993; 3:299-309.

31. Cockcroft VB, Ortells MO, Thomas P et al. Homologies and disparities of glutamate receptors: A critical analysis. Neurochem Int 1993; 23:583-594.
32. MacDonald RL, Olsen RW. $GABA_A$ receptor channels. Annu Rev Neurosci 1994; 17:569-602.
33. Hughes AL, Nei M. Pattern of nucleotide substitution at major histocompatibility complex class I loci reveals overdominant selection. Nature 1988; 335:167-170.
34. Ngai J, Dowling MM, Buck L et al. The family of genes encoding odorant receptors in the channel catfish. Cell 1993; 72:657-666.
35. Nei M. Molecular Evolutionary Genetics. Columbia University Press, New York, 1987.
36. Li WH. Unbiased estimation of the rates of synonymous and nonsynonymous substitution. J Mol Evol 1993; 36:96-99.
37. Løvtrup S. The phylogeny of vertebrata. John Wiley and Sons, 1977.
38. Ortells MO. A prediction of the secondary structure of the nicotinic acetylcholine receptor nontransmembrane regions. PROTEINS: Struc Func Genet (in press).
39. Couturier S, Bertrand D, Matter J et al. A neuronal nicotinic acetylcholine receptor subunit (α7) is developmentally regulated and forms a homo-oligomeric channel blocked by α-BTX. Neuron 1990; 5:847-856.
40. Marshall J, Buckingham SD, Shingai R et al. Sequence and functional expression of a single α subunit of an insect nicotinic acetylcholine receptor. EMBO J 1990; 9:4391-4398.

CHAPTER 3

The Ligand Binding Domains of the Nicotinic Acetylcholine Receptor

Richard J. Prince and Steven M. Sine

The concept of a specific binding site for nicotinic agonists dates back to the early physiological studies of Langley[1] and Dale.[2] During the intervening century, our understanding of the nicotinic receptor advanced roughly in step with key discoveries or technical advances. Development of microelectrode and single cell recording technology allowed definition of the activation and desensitization processes and the pore mechanism of ion transport. Discovery of the receptor-rich *Torpedo* electric organ and α-neurotoxins allowed biochemical analysis and definition of subunit composition and stoichiometry. Cloning of the subunit cDNAs revealed primary structure of each subunit, and with introduction of expression systems, allowed defined changes in structure. Introduction of the patch-clamp technique revealed switching of single receptors between functional states, and combined with mutagenesis, allowed identification of structural and functional domains. Our present view of three-dimensional structure owes to the ability to form two-dimensional crystal lattices of the receptor, suitable for analysis by modern electron diffraction methods. Though still not resolved at the atomic level, the three-dimensional structure of the receptor is beginning to emerge through combination of a range of experimental approaches. This chapter aims to draw together current knowledge of the acetylcholine receptor (AChR) ligand binding sites as revealed by structure-function and mutagenesis studies.

The Nicotinic Acetylcholine Receptor: Current Views and Future Trends,
edited by Francisco J. Barrantes.

Historical Overview

Biochemical Characterization

We hesitate to begin this chapter by repeating the cliché that the AChR is the best studied member of the cys-loop receptor superfamily. Tired as this phrase now is, it is also true. Biochemical and structural characterization of cys-loop receptors present several strategic problems. First, receptors represent only a small fraction of the protein in most tissues. Second, receptors are oligomeric membrane proteins whose function and structural integrity depend upon correct glycosylation, assembly and maintenance of a lipid environment. These factors weigh heavily against use of bacterial or yeast systems for expression of large amounts of receptor and make crystallization for X-ray diffraction extremely difficult. For the AChR, however, these problems were partly solved by discovery in the 1960s that the electric organ in the *Torpedo* ray is an extremely rich source of AChR.

Biochemical characterization of the AChR from *Torpedo* revealed several key features of receptor architecture. Analysis using polyacrylamide gel electrophoresis showed the receptor is composed of four distinct subunits, α (40 kDa), β (48 kDa), γ (58 kDa) and δ (64 kDa). Of these, only α appeared to bind site-directed labeling reagents, and the stoichiometry of acetylcholine (ACh) binding sites was one per 125 kDa of receptor mass.[3] Subsequently, sedimentation analysis of the intact receptor complex revealed a molecular mass of 250 kDa. Together these observations pointed to a subunit stoichiometry of 2αβγδ and two ligand binding sites per AChR.[4] Direct measurement of subunit stoichiometry was later achieved by N-terminal amino acid sequencing of all four subunits.[5] Reconstitution of purified receptor, containing only α, β, γ and δ subunits, demonstrated that the ion channel and the ACh binding sites are integral components of the receptor complex.[6,7]

The question of the order of the subunits around the central channel remains even today. However, three lines of evidence favor the arrangement (αγαδβ) shown in Figure 3.1B. First, electron micrographs of *Torpedo* AChR labeled with cobra α-toxin showed that the two α subunits are not adjacent to each other. Second, the location of the disulfide bond joining δ subunits of adjacent pentamers showed that δ is not the subunit between the two α subunits.[8] Similarly, specific crosslinking of β subunits in adjacent receptors using diamide revealed that β is also not the subunit between the α subunits.[9] Finally, expression studies omitting one or more subunits revealed high affinity binding sites formed by αγ and αδ subunit pairs, and further that the γ and δ subunits are functionally homologous in their ability to substitute for each other in the pentameric receptor.[10-12] Apposition of αγ and αδ subunit pairs was further demonstrated by photolabeling of α, γ, and δ subunits by *d*-tubocurarine (dTC)[13] and by specific crosslinking of α and δ subunits.[14] These findings, together with symmetrical positioning of each subunit in the pentamer, indicate that the subunit between the two α subunits is the γ subunit. There remains lack of consensus, however, as Kubalek et al[15] imaged receptors complexed with subunit-specific antibodies and concluded that the β subunit is the lone subunit between the two α subunits.

Overall Structure

Some aspects of AChR structure are reviewed in chapter 5 of this book (see also refs. 16 and 17). However, the following overview provides context for description of binding site architecture. Each of the four subunits possess an amino-terminal extracellular domain comprising approximately 50% of the AChR protein mass. Contained within this domain are the two agonist binding sites generated by apposition of the α with either the γ or the δ subunit. Also within the extracellular domain is the ubiquitous cys-loop, a disulfide bridged β-hairpin formed between C128 and C142, which is the most highly conserved domain in this receptor superfamily. Although the cys-loop is highly conserved, its functional significance is poorly understood. Mutagenesis studies showed that it does not contribute to the ligand binding site, but rather contributes to subunit assembly.[18-20] Carboxyl-terminal to the extracellular domain are four putative transmembrane domains (M1-M4), the second of which forms the lining of the ion-channel as a discontinuous α-helix (for reviews see refs. 16, 17 and 21). Figure 3.1 summarizes structu5re and subunit topology of the AChR.

N-terminal sequencing of the AChR subunits paved the way for the cloning of the subunit cDNAs.[5] Using degenerate oligonucleotides derived from sequencing, Numa's group cloned cDNAs encoding the α, β, γ and δ subunits.[22-24] Simultaneously, Ballivet et al[25] used hybrid-selection followed by immunoprecipitation with

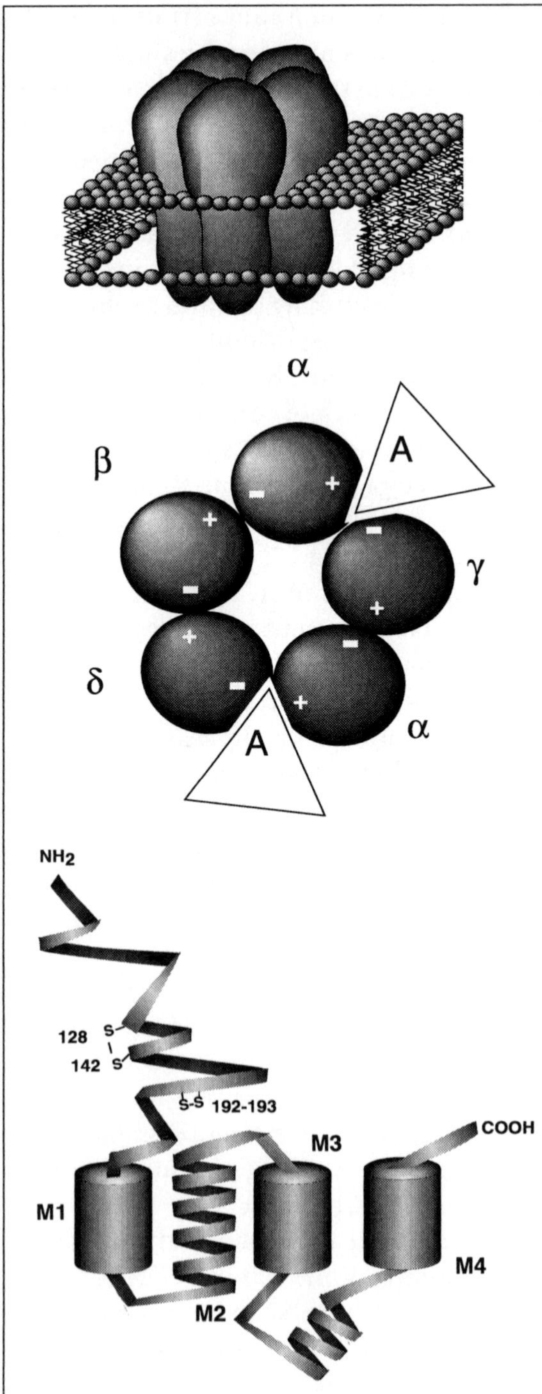

Fig. 3.1. Gross structural topology of the AChR complex.

(A) The five subunits are arrayed around the central ion channel like staves of a barrel forming a gated ion-conducting pathway across the cell membrane.

(B) The likely order of subunits around the ion channel is αγαδβ with the agonist binding sites formed at the αγ and αδ subunit interfaces. Each interface receives contributions from the (+) face of one subunit and the (–) face of its neighbor.

(C) Hydrophobicity analysis suggests that each subunit crosses the membrane four times with the N and C termini both being extracellular. M2, which likely forms the lining of the ion channel, is thought to be a kinked α-helix. Disulfide bonds are present between residues C128 and C142 in all subunits and in α subunits an additional disulfide is formed between residues C192 and C193.

subunit-specific antibodies to clone γ. Later studies revealed that AChR from adult muscle contains an ε subunit in place of the fetal γ subunit.[26] Cloning of electric organ subunits in turn laid the foundation for cloning muscle AChR subunits from a wide variety of species, as well as the related neuronal nicotinic receptor subunits (see refs. 27 and 28 for citation of cloning studies). Chapter 7 of this volume provides an overview of neuronal AChR structure and function.

Sequence comparisons of nicotinic AChR subunits with those from $GABA_A$, $5\text{-}HT_3$ and inhibitory glycine receptors reveal broader evolutionary relationships. Homology in primary and predicted secondary structures suggests that these receptors, together termed the cys-loop receptor or ligated-gated ion channel (LGIC)superfamily, evolved from a common ancestor (see chapter 2 in this volume).

The high degree of homology between AChR subunits, both in primary and predicted secondary structures, suggests that the polypeptide chains of each subunit fold into similar "basic scaffolds." This basic scaffold hypothesis predicts that residues equivalent in the linear sequence occupy equivalent positions in three dimensional space in each subunit. Differences between subunits thus owe mainly to differences in primary structure rather than to major differences in secondary or tertiary structures. Assuming each subunit orients around the central ion channel with the same handedness, the basic scaffold hypothesis predicts polarization of each subunit in terms of residues at each subunit interface. Thus each subunit interface is composed of a (+) face from one subunit and a (−) face from its neighbor (Fig. 3.1B), and these faces harbor residues equivalent in the linear sequences of the subunits.

AChR Ligands

Naturally occurring toxins from both plants and animals have been central to investigations of nicotinic receptor structure and function (Fig. 3.2A). Because of the vital role of the AChR in voluntary muscle, production of AChR-selective toxins is a common theme in animal hunting and defense strategies. Also many species of plants produce AChR active toxins to protect against consumption by animals and insects.

Perhaps the most renowned toxin targeted against the AChR is α-bungarotoxin (α-BTX) from the Taiwanese krait *Bungarus multicinctus*. This 74 amino acid peptide binds competitively to the muscle

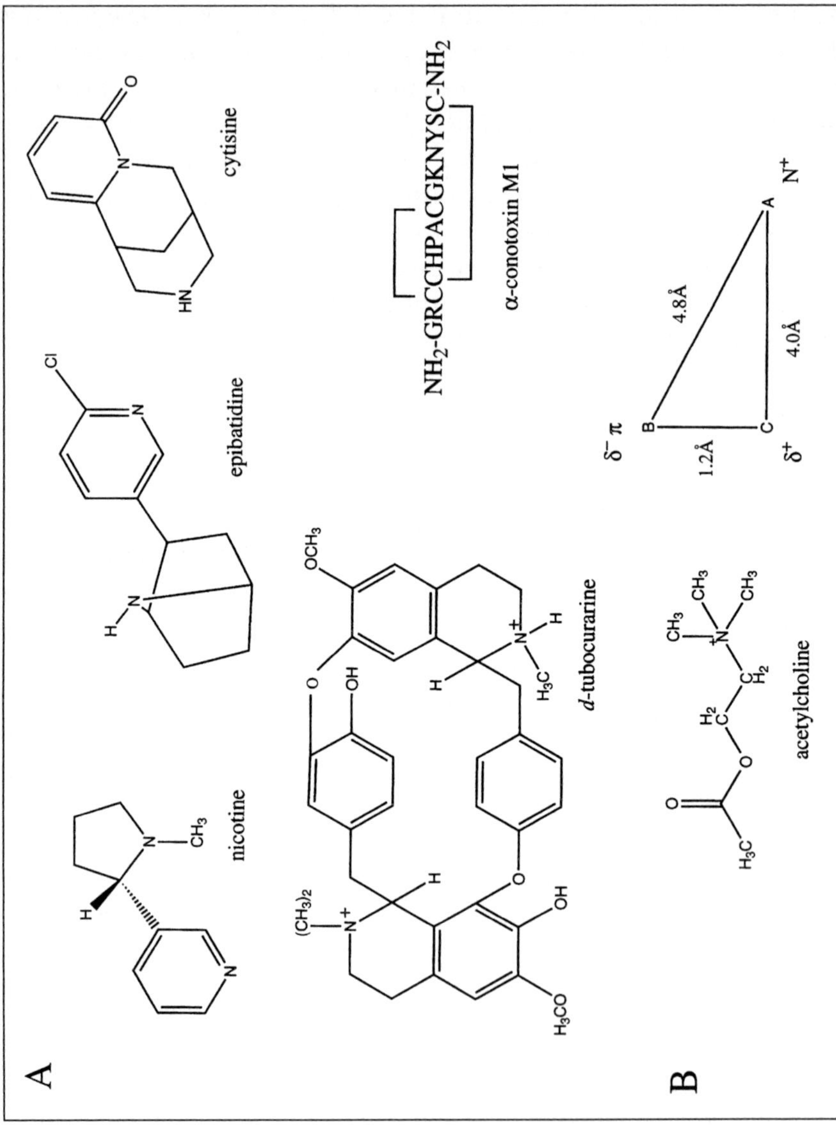

Fig. 3.2. (A) The structures of various naturally occurring nicotinic AChR ligands. (B) The structure of acetylcholine and the classic nicotinic receptor pharmacophore. In the pharmacophore triangle point A corresponds to the quaternary nitrogen of acetylcholine, B to the carbonyl oxygen and C to the carbonyl carbon. The pharmacophore structure was derived from consideration of a range of nicotinic agonists[34] whose structures constrain the interatomic distances shown.

AChR with subnanomolar affinity. As its binding is essentially irreversible, α-BTX has been of great utility in the labeling and purification of the receptor. A minor component of the venom of this snake, κ—or neuronal—bungarotoxin,[29] binds with high affinity to certain subtypes of neuronal AChR.

Another class of ligands used extensively to characterize the AChR are the curare alkaloids derived from plants of the Strychnos family. Naturally occurring curares, notorious for their use in South American arrow poisons, include tubocurarine and toxiferine. These compounds are competitive antagonists at the AChR and together with synthetic analogs such as alcuronium and dimethyl-tubocurarine, have found widespread use as muscle relaxants. Other compounds with curare-like actions, but which are chemically distinct from the curare alkaloids, include pancuronium, gallamine and compounds of the erythroidine family (see ref. 30 for a review of the pharmacology of curare, its synthetic analogs and curare-like compounds). A recent, significant addition to our pharmacological arsenal has been the alkaloid epibatidine, originally isolated from the skin of an Ecuadorian poison arrow frog, *Epipedobates tricolor*. Epibatidine binds with nanomolar to femtomolar affinity to several subtypes of neuronal AChRs.[31,32] Other important naturally occurring ligands include α-conotoxins from marine fish-hunting snails, the coral toxin lophotoxin, cytisine from *Laburnum* seeds and the tobacco alkaloid nicotine.

The minimal structure required for agonist activity is known classically as the pharmacophore, which specifies a three-dimensional arrangement of essential functional groups. Considering the structures of a range of agonists, several groups have proposed that the AChR pharmacophore consists of a three point interaction with the receptor (Fig. 3.2B).[33-35] Point A is a quaternary nitrogen, B is a π-bonded electronegative atom capable of hydrogen bonding, and point C is the positive end of the dipole set up by point B. Further, the distances between these points were constrained by the geometries of rigid ligands such as cytisine; however, the fact that the tetramethylammonium ion is an efficacious although low-affinity agonist demonstrates that the minimal AChR pharmacophore is actually a single point: a quaternary nitrogen. Points B and C in Figure 3.2, which in ACh correspond to the oxygen and carbon of the carbonyl group, can therefore be regarded as accessory anchor points, which confer high affinity agonist binding, but are not required for channel activation.

Receptor Activation and Ligand Binding

The AChR is designed to stay closed in the absence of ACh and to open its channel with high probability in its presence. Shuttling between open and closed states occurs without direct energy input and is driven by changes in noncovalent bonds between ACh and the binding site (see ref. 36 and chapter 4 in this volume for the treatment of the thermodynamics of ligand binding). Thus, ACh binds to the resting state of the AChR with low affinity, whereas it binds to the open channel and desensitized states with high affinity. Studies of the ligand binding site naturally led investigators to measure binding or dose-response relationships for agonists. However, such measurements are often made under conditions that include contributions from resting, open, and desensitized states, each of which binds agonist with different affinity. Thus *apparent* dissociation constants from such studies represent weighted averages of the affinity of each functional state. The dependence of apparent agonist affinity on functional state can be understood from the following description of activation and desensitization of the AChR:

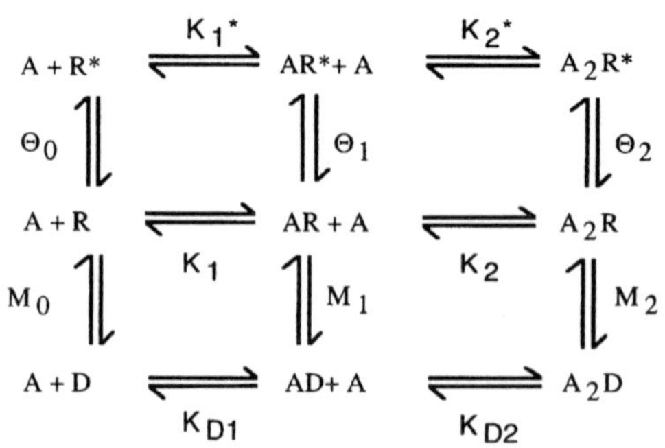

Scheme 1

In Scheme 1 the AChR assumes three functional states, closed (R), open (R*) and desensitized (D) (see chapter 6 and ref. 37 for review of nicotinic AChR function). θ_0 and M_0 are the equilibrium constants governing channel opening and desensitization; these are small numbers, minimizing these functional states in the absence of ACh. Channel opening and desensitization are driven by tighter binding of ACh to the R* and D states, that is, K_1 (the dissociation constant governing ACh binding) is larger than K_1^* and K_{D1}. Similarly, K_2 is greater than K_2^* and K_{D2}. Thus, sufficient energy to overcome the stable closed state is provided by ACh binding to two sites per receptor.[36] Inspection of Scheme 1 indicates that both binding and isomerization steps contribute to apparent affinity, so changes in both steps must be considered to interpret mutagenesis studies. To date, single channel kinetic analysis, which can resolve microscopic rate constants, is the only method that allows measurement of intrinsic affinities for agonist.

Contributions of the α-Subunits

Affinity Labeling Studies

Early affinity labeling studies using [^{3}H]MBTA (4-(N-maleimido)benzyltrimethyl-ammonium[3] identified the α subunit as a major component of the ACh binding site. Once primary sequences were known, residues labeled by affinity probes could be identified. The labeled residues were found to cluster into three linearly separate regions, leading to a three loop model of the contribution of the α subunit to the binding site (reviewed in ref. 38, see also Fig. 3.4). Loop A contains W86 and Y93 which are photolabeled by the competitive antagonist DDF [*p*-(*N,N*-dimethyl)]-aminobenzene-diazonium fluoroborate,[39] while Y93 is labeled by ACh mustard.[40] Loop B contains W149 and Y151, also labeled by DDF.[41] Loop C includes Y190, which is labeled by curare, DDF, nicotine and lophotoxin,[41-44] and Y198, which is labeled by curare, nicotine and DDF.[41,43,44] Included in loop C are C192 and C193, which form an unusual vicinal disulfide bond,[45] and are labeled by MBTA, DDF, nicotine and curare.[41,43,44,46]

Sequence Comparisons: Homology and Divergence

Residues identified by affinity labeling are conserved across all species and subtypes of nicotinic receptor α subunits (Fig. 3.3). One exception is the neuronal α_5 subunit which harbors Phe at position 93 and Asp at 190. Although α_5 does not form functional receptors when coexpressed with only one other neuronal subunit, it does when coexpressed with α_4 and β_2 subunits.[47] These heteromeric receptors differ in their activation and desensitization properties from receptors containing only α_4 and β_2, showing EC_{50}s markedly increased for ACh but decreased for nicotine. If α_5 contributes directly to the agonist binding site, the decrease in ACh sensitivity may be owing to the absence of Y93 and Y190. As discussed below, studies with muscle receptors containing mutations at αY93 and αY190 show decreased agonist affinity. However, a recent study by Wang et al,[48] in which α_5 coassembled with α_3 and either β_2 or β_4, revealed little change in

```
A1m   1   ------SEHETRLVAKLFED--YSSVVRPVEDHREIVQV
A1h   1   ------SEHETRLVAKLFKD--YSSVVRPVEDHRQVVEV
A1t   1   ------SEHETRLVANLLEN--YNKVIRPVEHHTHFVDI
A1c   1   ---------------------------------------
A1mg  1   ---------------------------------------
A1r   1   ------SEHETRLVAKLFKD--YSSVVRPVGDHREIVQV
A2r   1   QQGSHTHAEDRLFKHLFGG--YNRWARPVPNTSDVVIV
A3r   1   ------SEAEHRLFQYLFED--YNEIIRPVANVSHPVII
A4r   1   -IETRAHAEERLLKRLFSG--YNKWSRPVGNISDVVLV
A5r   1   ------AKHEDSLFRDLFED--YERWVRPVEHLSDKIKI
A6r   1   ------CVSEEQLFHTLFAH--YNRFIRPVENVSDPVTV
A7r   1   ------GEFQRRLYKELVKN--YNPLERPVANDSQPLTV
A8c   1   ------GESQRRLYRDLLRN--YNRLERPVMNDSQPIVV
A9r   1   VETANGKYAQKLFSDLFED--YSSALRPVEDTDAVLNV
Gm    1   ------RNQEERLLADLMRN--YDPHLRPAERDSDVVNV
Gh    1   ------RNQEERLLADLMQN--YDPNLRPAERDSDVVNV
Dm    1   ------LNEEQRLIQHLFNEKGYDKDLRPVARKEDKVDV
Dh    1   ------LNEEERLIRHLFQEKGYNKELRPVAHKEESVDV
Em    1   ------KNEELSLYHHLFDN--YDPECRPVRRPEDTVTI
Eh    1   ------KNEELRLYHHLFNN--YDPGSRPVREPEDTVTI
B1m   1   ------SEAEGQLIKKLFSN--YDSSVRPAREVGDRVGV
B1h   1   ------SEAEGRLREKLFSG--YDSSVRPAREVGDRVRV
B2h   1   ------TDTEERLVEHLLDPSRYNKLIRPATNGSELVTV
B3h   1   ----------------EWKG--YQKWVRPVLHSNDTIKV
B4h   1   ------------------------------TSSSQLISI

A1m   32  TVGLQLIQLINVDEVNQIVTTNVRLKQQWVDYNLKWNP
A1h   32  TVGLQLIQLINVDEVNQIVTTNVRLKQQWVDYNLKWNP
A1t   32  TVGLQLIQLISVDEVNQIVETNVRLRQQWIDVRLRWNP
A1c   1   --------------------------------------
A1mg  1   --------------------------------------
A1r   32  TVGLQLIQLINVDEVNQIVTTNVRLKQQWVDYNLKWNP
A2r   37  RFGLSIAQLIDVDEKNQMMTTNVWLKQEWNDYNVRWDP
A3r   32  QFEVSMSQLVKVDEVNQIMETNLWLKQIWNDYKLKWKP
A4r   36  RFGLSIAQLIDVDEKNQMMTTNVWVKQEWHDYKLRWDP
A5r   32  KFGLAISQLVDVDEKNQLMTTNVWLKQEWIDVKLRWNP
A6r   32  HFELAITQLANVDEVNQIMETNLWLRHVWKDYRLCWDP
A7r   32  YFSLSLLQIMDVDEKNQVLTTNIWLQMSWTDHYLQWNM
A8c   32  ELQLSLLQIIDVDEKNQVLITNAWLQMYWVDIYLSWDQ
A9r   37  TLQVTLSQIKDMDERNQILTAYLWIRQTWHDAYLTWDR
Gm    32  SLKLTLTNLISLNEREEALTTNVWIEMQWCDYRLRWDP
Gh    32  SLKLTLTNLISLNEREEALTTNVWIEMQWCDYRLRWDP
Dm    34  ALSLTLSNLISLKEVEETLTTNVWIDHAWVDSRLQWDA
Dh    34  ALALTLSNLISLKEVEETLTTNVWIEHGWTDNRLKWNA
Em    32  TLKVTLTNLISLNEKEETLTTSVWIGIDWHDYRLNYSK
Eh    32  SLKVTLTNLISLNEKEETLTTSVWIGIDWQDYRLNYSK
B1m   32  SIGLTLAQLISLNEKDEEMSTKVYLDLEWTDYRLSWDP
B1h   32  SVGLILAQLISLNEKDEEMSTKVYLDLEWTDYRLSWDP
B2h   34  QLMVSLAQLISVHEREQIMTTNVWLTQEWEDYRLTWKP
B3h   22  YFGLKISQLVDVDEKNQLMTTNVWLKQEWTDHKLRWNP
B4h   10  KLQLSLAQLISVNERDEIMTTNVWLKQEWTDYRLTWNS
             n                   n n
```

A

Fig. 3.3A,B,C. Alignment of nicotinic AChR subunit extracellular domains. Residues conserved across at least 50% of the analyzed sequences are boxed. Annotations above and below the text indicate residues implicated in binding site formation as follows: α, alpha residues; n, non-α residues; underlined, identified by mutagenesis; italic, identified by affinity labeling; underlined and italic, identified by both affinity labeling and mutagenesis. Sequence names are abbreviated as follows: A, α; B, β; G, γ; D, δ; E, ε; m, mouse; h, human; t, *Torpedo*; r, rat; mg, mongoose; A1c, cobra α1; A8c, chicken α8. Numbering is based on known sequence. Mongoose and cobra α1 are both partial sequences.

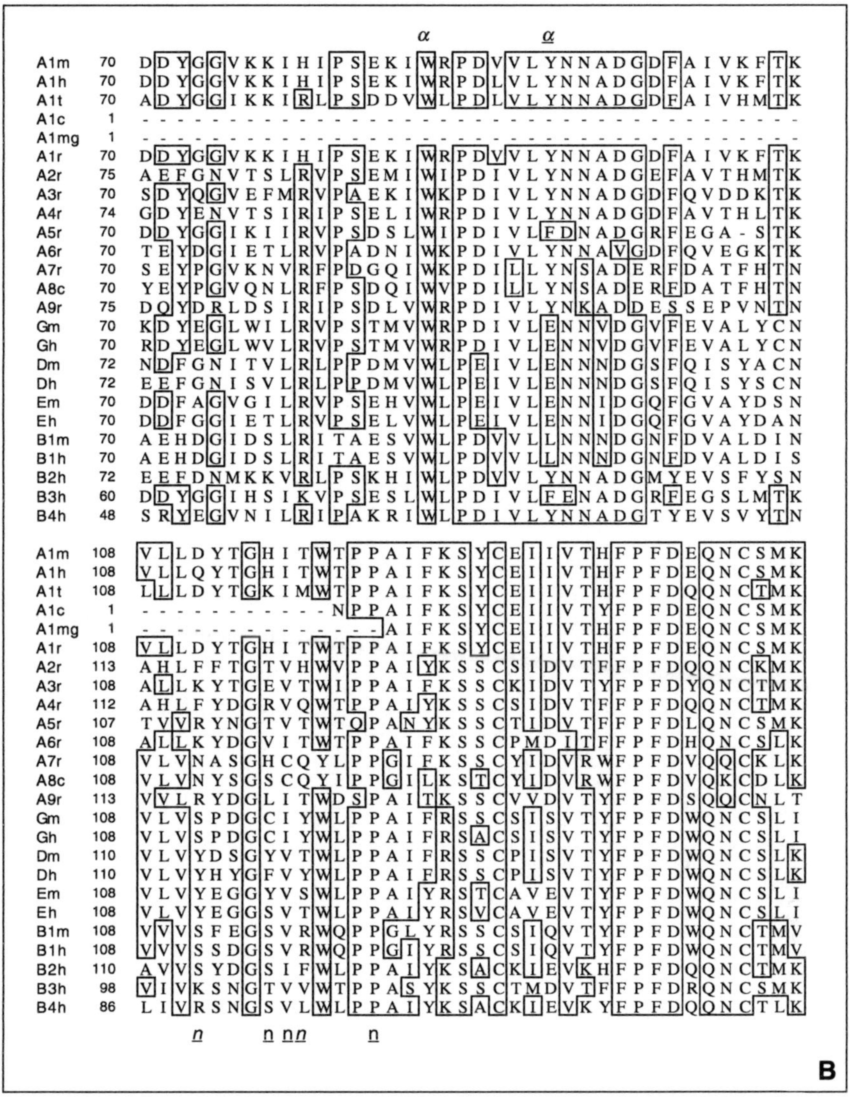
α α
A1m 70 D D Y G G V K K I H I P S E K I W R P D V V L Y N N A D G D F A I V K F T K
A1h 70 D D Y G G V K K I H I P S E K I W R P D L V L Y N N A D G D F A I V K F T K
A1t 70 A D Y G G I K K I R L P S D D V W L P D L V L Y N N A D G D F A I V H M T K
A1c 1 -
A1mg 1 -
A1r 70 D D Y G G V K K I H I P S E K I W R P D V V L Y N N A D G D F A I V K F T K
A2r 75 A E F G N V T S L R V P S E M I W I P D I V L Y N N A D G E F A V T H M T K
A3r 70 S D Y Q G V E F M R V P A E K I W K P D I V L Y N N A D G D F Q V D D K T K
A4r 74 G D Y E N V T S I R I P S E L I W R P D I V L Y N N A D G D F A V T H L T K
A5r 70 D D Y G G I K I I R V P S D S L W I P D I V L F D N A D G R F E G A - S T K
A6r 70 T E Y D G I E T L R V P A D N I W K P D I V L Y N N A V G D F Q V E G K T K
A7r 70 S E Y P G V K N V R F P D G Q I W K P D I L L Y N S A D E R F D A T F H T N
A8c 70 Y E Y P G V Q N L R F P S D Q I W V P D I L L Y N S A D E R F D A T F H T N
A9r 75 D Q Y D R L D S I R I P S D L V W R P D I V L Y N K A D D E S S E P V N T N
Gm 70 K D Y E G L W I L R V P S T M V W R P D I V L E N N V D G V F E V A L Y C N
Gh 70 R D Y E G L W V L R V P S T M V W R P D I V L E N N V D G V F E V A L Y C N
Dm 72 N D F G N I T V L R L P P D M V W L P E I V L E N N N D G S F Q I S Y A C N
Dh 72 E E F G N I S V L R L P P D M V W L P E I V L E N N N D G S F Q I S Y S C N
Em 70 D D F A G V G I L R V P S E H V W L P E I V L E N N I D G Q F G V A Y D S N
Eh 70 D D F G G I E T L R V P S E L V W L P E I V L E N N I D G Q F G V A Y D A N
B1m 70 A E H D G I D S L R I T A E S V W L P D V V L L N N N D G N F D V A L D I N
B1h 70 A E H D G I D S L R I T A E S V W L P D V V L L N N N D G N F D V A L D I S
B2h 72 E E F D N M K K V R L P S K H I W L P D V V L Y N N A D G M Y E V S F Y S N
B3h 60 D D Y G G I H S I K V P S E S L W L P D I V L F E N A D G R F E G S L M T K
B4h 48 S R Y E G V N I L R I P A K R I W L P D I V L Y N N A D G T Y E V S V Y T N
A1m 108 V L L D Y T G H I T W T P P A I F K S Y C E I I V T H F P F D E Q N C S M K
A1h 108 V L L Q Y T G H I T W T P P A I F K S Y C E I I V T H F P F D E Q N C S M K
A1t 108 L L L D Y T G K I M W T P P A I F K S Y C E I I V T H F P F D Q Q N C T M K
A1c 1 - - - - - - - - - - - N P P A I F K S Y C E I I V T Y F P F D E Q N C S M K
A1mg 1 - - - - - - - - - - - - - - A I F K S Y C E I I V T H F P F D E Q N C S M K
A1r 108 V L L D Y T G H I T W T P P A I F K S Y C E I I V T H F P F D E Q N C S M K
A2r 113 A H L F F T G T V H W V P P A I Y K S S C S I D V T F F P F D Q Q N C K M K
A3r 108 A L L K Y T G E V T W I P P A I F K S S C K I D V T Y F P F D Y Q N C T M K
A4r 112 A H L F Y D G R V Q W T P P A I Y K S S C S I D V T F F P F D Q Q N C T M K
A5r 107 T V V R Y N G T V T W T Q P A N Y K S S C T I D V T F F P F D L Q N C S M K
A6r 108 A L L K Y D G V I T W T P P A I F K S S C P M D I T F F P F D H Q N C S L K
A7r 108 V L V N A S G H C Q Y L P P G I F K S S C Y I D V R W F P F D V Q Q C K L K
A8c 108 V L V N Y S G S C Q Y I P P G I L K S T C Y I D V R W F P F D V Q K C D L K
A9r 113 V V L R Y D G L I T W D S P A I T K S S C V V D V T Y F P F D S Q Q C N L T
Gm 108 V L V S P D G C I Y W L P P A I F R S S C S I S V T Y F P F D W Q N C S L I
Gh 108 V L V S P D G C I Y W L P P A I F R S A C S I S V T Y F P F D W Q N C S L I
Dm 110 V L V Y D S G Y V T W L P P A I F R S S C P I S V T Y F P F D W Q N C S L K
Dh 110 V L V Y H Y G F V Y W L P P A I F R S S C P I S V T Y F P F D W Q N C S L K
Em 108 V L V Y E G G Y V S W L P P A I Y R S T C A V E V T Y F P F D W Q N C S L I
Eh 108 V L V Y E G G S V T W L P P A I Y R S V C A V E V T Y F P F D W Q N C S L I
B1m 108 V V V S F E G S V R W Q P P G L Y R S S C S I Q V T Y F P F D W Q N C T M V
B1h 108 V V V S S D G S V R W Q P P G I Y R S S C S I Q V T Y F P F D W Q N C T M V
B2h 110 A V V S Y D G S I F W L P P A I Y K S A C K I E V K H F P F D Q Q N C T M K
B3h 98 V I V K S N G T V V W T P P A S Y K S S C T M D V T F F P F D R Q N C S M K
B4h 86 L I V R S N G S V L W L P P A I Y K S A C K I E V K Y F P F D Q Q N C T L K
n n n n n
B

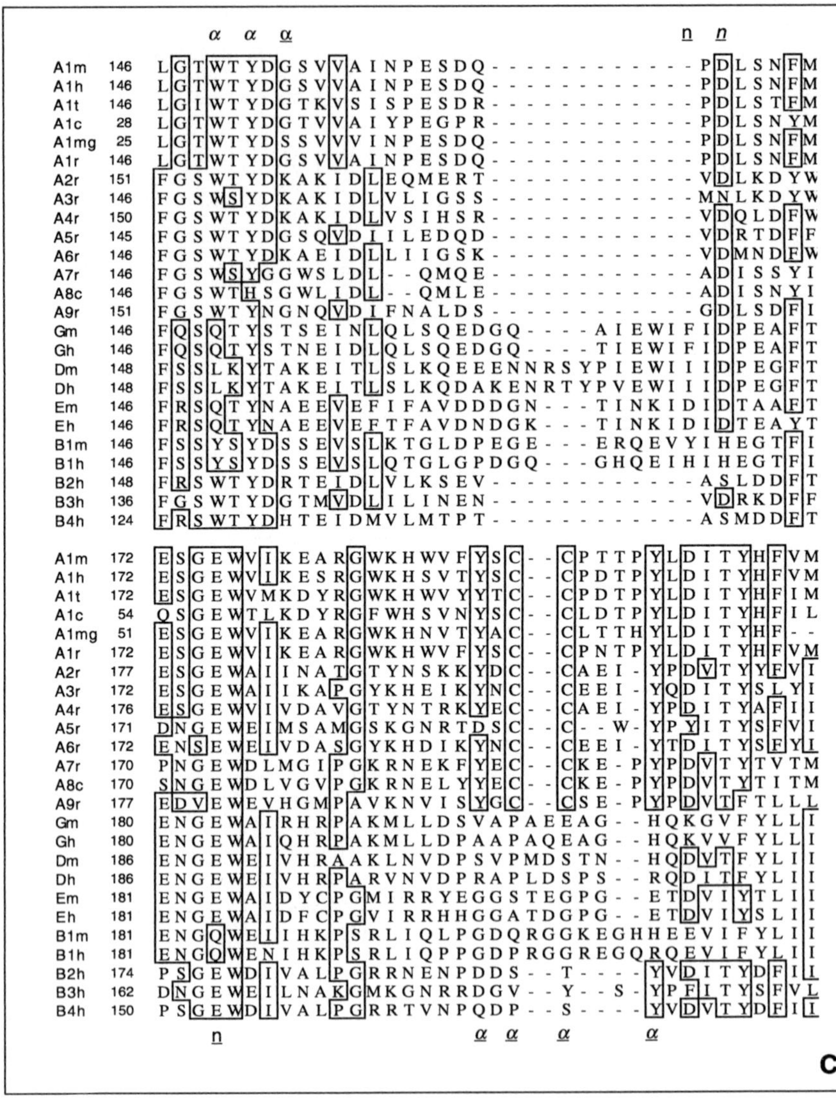
α α α n n
A1m 146 L G T W T Y D G S V V A I N P E S D Q - - - - - - - - - - - - - P D L S N F M
A1h 146 L G T W T Y D G S V V A I N P E S D Q - - - - - - - - - - - - - P D L S N F M
A1t 146 L G I W T Y D G T K V S I S P E S D R - - - - - - - - - - - - - P D L S T F M
A1c 28 L G T W T Y D G T V V A I Y P E G P R - - - - - - - - - - - - - P D L S N Y M
A1mg 25 L G T W T Y D S S V V V I N P E S D Q - - - - - - - - - - - - - P D L S N F M
A1r 146 L G T W T Y D G S V V A I N P E S D Q - - - - - - - - - - - - - P D L S N F M
A2r 151 F G S W T Y D K A K I D L E Q M E R T - - - - - - - - - - - - - V D L K D Y W
A3r 146 F G S W S Y D K A K I D L V L I G S S - - - - - - - - - - - - - M N L K D Y W
A4r 150 F G S W T Y D K A K I D L V S I H S R - - - - - - - - - - - - - V D Q L D F W
A5r 145 F G S W T Y D G S Q V D I I L E D Q D - - - - - - - - - - - - - V D R T D F F
A6r 146 F G S W T Y D K A E I D L L I I G S K - - - - - - - - - - - - - V D M N D F W
A7r 146 F G S W S Y G G W S L D L - - Q M Q E - - - - - - - - - - - - - A D I S S Y I
A8c 146 F G S W T H S G W L I D L - - Q M L E - - - - - - - - - - - - - A D I S N Y I
A9r 151 F G S W T Y N G N Q V D I F N A L D S - - - - - - - - - - - - - G D L S D F I
Gm 146 F Q S Q T Y S T S E I N L Q L S Q E D G Q - - - - A I E W I F I D P E A F T
Gh 146 F Q S Q T Y S T N E I D L Q L S Q E D G Q - - - - T I E W I F I D P E A F T
Dm 148 F S S L K Y T A K E I T L S L K Q E E E N N R S Y P I E W I I I D P E G F T
Dh 148 F S S L K Y T A K E I T L S L K Q D A K E N R T Y P V E W I I I D P E G F T
Em 146 F R S Q T Y N A E E V E F I F A V D D D G N - - - T I N K I D I D T A A F T
Eh 146 F R S Q T Y N A E E V E F T F A V D N D G K - - - T I N K I D I D T E A Y T
B1m 146 F S S Y S Y D S S E V S L K T G L D P E G E - - - E R Q E V Y I H E G T F I
B1h 146 F S S Y S Y D S S E V S L Q T G L G P D G Q - - - G H Q E I H I H E G T F I
B2h 148 F R S W T Y D R T E I D L V L K S E V - - - - - - - - - - - - - A S L D D F T
B3h 136 F G S W T Y D G T M V D L I L I N E N - - - - - - - - - - - - - V D R K D F F
B4h 124 F R S W T Y D H T E I D M V L M T P T - - - - - - - - - - - - - A S M D D F T
A1m 172 E S G E W V I K E A R G W K H W V F Y S C - - C P T T P Y L D I T Y H F V M
A1h 172 E S G E W V I K E S R G W K H S V T Y S C - - C P D T P Y L D I T Y H F V M
A1t 172 E S G E W V M K D Y R G W K H W V Y Y T C - - C P D T P Y L D I T Y H F I M
A1c 54 Q S G E W T L K D Y R G F W H S V N Y S C - - C L D T P Y L D I T Y H F I L
A1mg 51 E S G E W V I K E A R G W K H N V T Y A C - - C L T T H Y L D I T Y H F - -
A1r 172 E S G E W V I K E A R G W K H W V F Y S C - - C P N T P Y L D I T Y H F V M
A2r 177 E S G E W A I I N A T G T Y N S K K Y D C - - C A E I - Y P D V T Y Y F V I
A3r 172 E S G E W A I I K A P G Y K H E I K Y N C - - C E E I - Y Q D I T Y S L Y I
A4r 176 E S G E W V I V D A V G T Y N T R K Y E C - - C A E I - Y P D I T Y A F I I
A5r 171 D N G E W E I M S A M G S K G N R T D S C - - C - - W - Y P Y I T Y S F V I
A6r 172 E N S E W E I V D A S G Y K H D I K Y N C - - C E E I - Y T D I T Y S F Y I
A7r 170 P N G E W D L M G I P G K R N E K F Y E C - - C K E - P Y P D V T Y T V T M
A8c 170 S N G E W D L V G V P G K R N E L Y Y E C - - C K E - P Y P D V T Y T I T M
A9r 177 E D V E W E V H G M P A V K N V I S Y G C - - C S E - P Y P D V T F T L L L
Gm 180 E N G E W A I R H R P A K M L L D S V A P A E E A G - - H Q K G V F Y L L I
Gh 180 E N G E W A I Q H R P A K M L L D P A A P A Q E A G - - H Q K V V F Y L L I
Dm 186 E N G E W E I V H R A A K L N V D P S V P M D S T N - - H Q D V T F Y L I I
Dh 186 E N G E W E I V H R P A R V N V D P R A P L D S P S - - R Q D I T F Y L I I
Em 181 E N G E W A I D Y C P G M I R R Y E G G S T E G P G - - E T D V I Y T L I I
Eh 181 E N G E W A I D F C P G V I R R H H G G A T D G P G - - E T D V I Y S L I I
B1m 181 E N G Q W E L I H K P S R L I Q L P G D Q R G G K E G H H E E V I F Y L I I
B1h 181 E N G Q W E N I H K P S R L I Q P P G D P R G G R E G Q R Q E V I F Y L I I
B2h 174 P S G E W D I V A L P G R R N E N P D D S - - T - - - - Y V D I T Y D F I I
B3h 162 D N G E W E I L N A K G M K G N R R D G V - - Y - - S - Y P F I T Y S F V L
B4h 150 P S G E W D I V A L P G R R T V N P Q D P - - S - - - - Y V D V T Y D F I I
n α α α α
C

agonist sensitivity for $\alpha_3\alpha_5\beta_4$ receptors but increased sensitivity to both ACh and nicotine for $\alpha_3\alpha_5\beta_2$ receptors. These authors propose that α_5 occupies a position equivalent to that of β_1 in the muscle receptor and indirectly affects the affinity of $\alpha_3\alpha_5\beta_2$ receptors. Experiments designed to delineate the contribution of α_5 to the binding site may shed further light on the contribution of conserved residues in loops A and C.

Since sequence homology is usually very high across species (Fig. 3.3 and see chapter 2 in this volume), functional differences among species can often be traced to discrete amino acid differences. Cobra and mongoose AChRs, for example, are highly resistant to α-BTX. As cobra venom contains a peptide homologous to α-BTX, and the mongoose is a predator of the cobra, this resistance is of significant evolutionary value to both animals. By comparing sequences of mongoose and cobra α-subunits with α-subunits from α-BTX sensitive species, Kreienkamp et al[49] identified a series of residues as candidate determinants of α-BTX sensitivity. Subsequent point mutations demonstrated that N-linked glycosylation sites in loop C, at position 187 in mongoose and 189 in cobra, were sufficient to inhibit the binding of toxin without significantly altering affinities for agonists or small antagonists. Proximity of these residues to conserved residues in loop C suggests that insensitivity to α-BTX owes to steric occlusion by the bulky oligosaccharide.

Further insight into the contributions of the α-subunits to the ACh binding sites has come from identification and functional characterization of mutations underlying congenital myasthenic syndromes, myasthenia gravis-like disorders of nonautoimmune origin (see chapter 8 in this book for the treatment of molecular pathologies of the AChR). In one case, pathogenicity was traced to the mutation αG153S, which increased affinity of ACh for the resting closed state, and prolonged activation episodes by allowing multiple reopenings before ACh could dissociate.[50] αG153S is in loop B and may increase agonist affinity by affecting the contributions of W149 and Y151.

Mutagenesis Studies

The functional significance of residues identified by affinity labeling studies has been extensively investigated by site directed mutagenesis. The results show that contributions to affinity depend on whether the test ligand is an agonist or an antagonist, even though binding of these two classes of ligands is mutually exclusive. The different contributions may indicate different orientations in the binding site, or contributions to equilibria between functional states that indirectly affect affinity.

In one of the earliest studies, Mishina et al[51] converted C192 and C193 to serines and observed decreases in both α-bungarotoxin and ACh sensitivity. Contributions of the conserved aromatic residues Y93, W149, Y190 and Y198 were first investigated by dose response measurements. The mutations αY190F and αY198F markedly increased the EC_{50} for ACh, while mutation of other nearby conserved residues was largely without effect.[52] Similarly, in the homomeric receptor formed by the neuronal α7 subunit, mutation to Phe of Y92, W148 and Y187 (equivalent to Y93, W149 and Y190 of the muscle receptor) increased the EC_{50} for ACh.[53] To determine the mechanism by which Y93, Y190 and Y198 interact with agonist, Sine et al[54] mutated each tyrosine to serine and found roughly equivalent contributions to both ACh and tetramethylammonium (TMA) affinity. The similar effects on ACh and TMA suggest that these residues stabilize the quaternary ammonium group of ACh. Introducing different side chains, however, revealed fundamentally different contributions of these residues; aromaticity is essential for Y198, whereas the aromatic hydroxyl is required for Y93 and Y190. Nowak et al[55] used a novel stop codon suppresser tRNA method to engineer unnatural amino acids into positions α93, 190 and 198; these authors also found that aromaticity is essential for Y198, whereas the aromatic hydroxyl is essential for Y93. Further, ring substitution of Y190 (e.g., 2-fluorotyrosine) completely abolished responses to ACh, while O-substitution markedly decreased ACh sensitivity.

Studies of competitive antagonists revealed contributions distinct from agonists for Y93, Y190 and Y198.[54] Mutation of Y93 did not affect dimethyl-*d*-tubocurarine (DMT) affinity, while mutations of Y190 and Y198 showed novel side chain dependencies for DMT. Y198F, for example, enhanced DMT affinity in contrast to the decreases in ACh and TMA affinities. Studies of side chain specificity showed that Y198 stabilizes DMT through aromatic/quaternary interactions.[77] A similar effect of Y198F on the binding of the unmethylated analog dTC was noted by Filatov et al.[56]

Although binding and macroscopic dose-response measurements indicate that the conserved tyrosines stabilize agonist, such measurements do not reveal the microscopic rate constant affected by the mutations (see Scheme 1). To determine the rate constants affected by αY93F and αY198F, Aylwin and White[57] analyzed the kinetics of single channel currents. They found that both mutations decreased the intrinsic affinities of the binding sites but did not significantly affect channel gating rate constants. Both mutations primarily slowed the association rate of ACh. Chen et al[58] found that αY190F decreases agonist affinity, again by decreasing the association rate of ACh. Paradoxically, Y190F decreases rates of agonist dissociation, changes opposite to those expected from the overall decrease of affinity. These results suggest that Y190 lies along the diffusion pathway which leads to the primary docking site, forming a barrier to entry and escape of agonist. Also, approximately half of the decrease in apparent affinity arose from slowing of the channel opening rate constant, suggesting that Y190 contributes to the activated complex between closed and open states.

The overall results suggest that Y93, Y190, and Y198 facilitate diffusion of ACh into the binding cleft. Evidence for a long diffusion pathway comes from Unwin's cryoelectron microscope studies which showed a narrow cleft formed by three apparently helical rods in the center of each α subunit.[59] By analogy to acetylcholinesterase, which contains an "aromatic gorge" leading to the ACh docking site 20 Å from the enzyme surface,[60] the cleft in the receptor was proposed to lead to the primary ACh binding site. The conserved aromatic residues W86, Y93, W149, Y151, Y190 and Y198 could line this cleft, acting as "molecular grease" to draw ACh from extracellular space into its primary binding site.

In summary, mutagenesis and affinity labeling studies reveal contributions of residues from three distinct regions of the α subunits to binding of agonists or antagonists. Each loop contains conserved aromatic residues which appear to stabilize ACh through either aromatic/quaternary or phenolate/quaternary interactions. Measurements of agonist binding kinetics together with data from structural studies suggests, however, that these conserved aromatic residues may not contribute to the primary docking site for ACh but rather may line an "aromatic gorge" through which ACh traverses before docking to as yet unidentified residues.

Contributions of the Non-α Subunits

As discussed above, early studies of AChR binding sites largely revealed contributions of the α subunits. However, a large body of evidence demonstrates that the two ligand binding sites are not identical. As the two α subunits are encoded by a single gene, and are thus identical in primary sequence, nonequivalence of the two binding sites is due to different contributions of the neighboring non-α subunits; one α is bounded by β and γ while the other is bounded by γ and δ (Fig. 3.1B). Early evidence for nonequivalent sites came from affinity labeling studies with [^{3}H]-MBTA and [^{3}H]-bromoacetylcholine[61,62] which demonstrated preferential incorporation of each into one of the two α subunits. Curariform antagonists also exhibited distinct affinities for the two sites,[63,64] and the different affinities were not due to negatively cooperative binding.[64] Subsequently, single channel kinetic analysis revealed distinct ACh affinities of the two binding sites in *Torpedo* and fetal mouse receptors.[65,66] A structurally diverse range of agonists and antagonists are now known to distinguish between the two sites, including α-conotoxin M1, lophotoxin, and epibatidine.[12,49,67]

Expression Studies

Given that the AChR binding sites are generated by α and non-α subunits, the natural question is what non-α subunits contribute to the sites. Blount and Merlie[11,68] addressed this question by constructing fibroblast cell lines stably expressing various pairs of subunits. Cell lines expressing α plus β, or even α alone, acquired α-bungarotoxin binding sites in a time dependent fashion, but these binding sites exhibited low affinity for ACh and curare. In contrast, cells expressing either α and δ or α and γ formed α-bungarotoxin binding sites with high affinity for agonists and antagonists. Moreover, the relative affinities of the αγ and αδ complexes for curare were close to those of the two binding sites in the native receptor. Sine and Claudio[12] made cell lines expressing α, β and γ or α, β and δ and found that each combination formed cell surface pentamers with compositions $\alpha_2\beta\gamma_2$ or $\alpha_2\beta\delta_2$, respectively. Both of these pentamers exhibited a single class of binding sites for DMT, with the affinity of $\alpha_2\beta\gamma_2$ corresponding to the high affinity site in the native pentamer, and the affinity of $\alpha_2\beta\delta_2$ corresponding to the low affinity site. These "triplet" receptors exhibited the correct selectivities for lophotoxin and α-conotoxin M1, but these ligands, like agonist, bound to $\alpha_2\beta\delta_2$ receptors with high affinity and $\alpha_2\beta\gamma_2$ receptors with low affinity.

Overall, these studies show that the binding sites are generated by αγ and αδ subunit pairs. Further, the γ and δ subunits can substitute for one another functionally, suggesting similar structural contributions to the two binding sites. Thus, ligand selectivity of the two binding sites of the AChR is likely determined by discrete differences in primary structure between the (–) faces of the γ or δ subunits, as postulated in the basic scaffold hypothesis.

Studies of neuronal nicotinic receptors also show that non-α subunits contribute to the binding sites. Luetje and coworkers[69-71] found that coexpression of α_2, α_3 or α_4 with either β_2 or β_4 yielded receptors with different sensitivities to both agonists and antagonists. For example, $\alpha_3\beta_2$ receptors are more sensitive to block by neuronal bungarotoxin than $\alpha_3\beta_4$ receptors. In contrast, $\alpha_2\beta_2$ receptors are insensitive to blockade by neuronal bungarotoxin but are much more sensitive to nicotine than $\alpha_3\beta_2$ receptors. These results too indicate that agonist binding sites in neuronal nicotinic receptors are generated by α and β subunits, and further, that the β_2 and β_4 subunits contribute residues equivalent to or neighboring selectivity determinants in the γ and δ subunits of muscle receptors.

The three sub-types of neuronal nicotinic AChR subunit, α_7, α_8 and α_9 form homomeric receptors when heterologously expressed. As these receptors contain only one type of subunit, either the agonist binding sites in such homomers are contained solely within the (+) face of the subunits or a single type of subunit contributes both the (+) and (–) faces of the ligand binding site; the latter possibility suggests that binding sites formed by α_7, α_8 or α_9 subunits contain residues equivalent to those in α_1 and in γ, δ or ε. Evolutionary relationships between nicotinic receptor subunits place α_7 and α_8 in the subfamily which diverged from a primitive ancestral AChR subunit before the main α family (containing α_1-α_6 and β_3) diverged from the main non-α family (containing β_1, β_2, β_4, δ, γ and ε) (see ref. 27 and chapter 2 of this book). Further, relative to the other subtypes, α_7 and α_8 appear to have diverged less from the putative common ancestor of the AChR family. α_7 and α_8 can thus be regarded as equally related to the main α and non-α subtypes, suggesting that homomer forming subunits contain determinants for both faces of the ligand binding site, a hypothesis recently supported by mutagenesis studies (see below).

Affinity Labeling

Reports of affinity labeling of non-α subunits are not as common as those of labeling the α subunit, and even fewer of these revealed the labeled residues. Nevertheless, affinity labeling studies strongly support the notion of interfacial binding sites containing contributions from both α and non-α subunits. Using α-bungarotoxin as a photolabel, Oswald and Changeux[72] demonstrated labeling of both the γ and δ subunits, while DDF photolabeled the γ subunit in an agonist-protectable manner.[73] Pedersen and Cohen[13] observed photolabeling of α, γ and δ subunits by [^{3}H]dTC, and half maximal incorporation into γ and δ occurred at dTC concentrations close to the K_ds of the respective high or low affinity sites. Subsequent photolabeling studies with [^{3}H]dTC suggest that the major incorporation sites for this ligand are the conserved tryptophans W55 in γ and W57 in δ.[44,74] Minor labeling sites were also detected at γY111 and γY117.[44] Czajkowski and Karlin[75] tethered the bifunctional crosslinking agent S-(2-[^{3}H]glycylamidoethyl)dithio-2-pyridine (GCP) to the reduced αC192-C193, and observed agonist-protectable incorporation of the carboxyl-specific end of GCP into a segment encompassing residues 164 to 257 of the of the δ subunit. This region of *Torpedo* δ harbors 11 aspartate or glutamate residues in the portion predicted to be extracellular. As the extended conformation of GCP is around 9 Å, one or more of these acidic residues lies within 9 Å of αC192-C193. Subsequent microsequencing identified δD165, δD180 and δE182 as GCP labeled residues.[14]

Mutagenesis/Structure Function Studies

Using the principle that homologous proteins contain transplantable domains that modify function, Sine[76] constructed γ/δ subunit chimeras to localize residues on the (–) face of these subunits that determine selectivity of DMT for the two binding sites. Three pairs of residues were identified, γI116/δV118, γY117/δT119 and γS161/δK163, that fully account for the higher affinity of the αγ relative to the αδ site. Further, the results supported the basic scaffold hypothesis, because mutation of the three residues in γ to their δ counterparts produced pure δ affinity. The determinant contributing most to high affinity of the αγ site, γY117, was found to stabilize DMT through quaternary/aromatic interactions, and αY198 on the (+) face of the site provided additional stabilization through symmetrical quaternary/aromatic interactions.[77] This study further demonstrated independent contributions of αY198 and γY117, suggesting that the two quaternary nitrogens in DMT bridge between γY117 and αY198.

Sine et al[78] also constructed γ/δ chimeras to identify residues that confer the ~10,000-fold higher affinity of conotoxin M1 for the αδ over the αγ site. Again, three pairs of residues, δS36/γK34, δY113/γS111 and δI178/γF172, were identified, each distant in the linear sequence. The findings again supported the basic scaffold hypothesis because mutation of the three determinants in δ to their γ counterparts converted conotoxin affinity to that conferred by the γ subunit. Interaction between δS36/γK34 and δI178/γF172 was also observed, both in dictating conotoxin affinity and expression levels of surface pentamers, suggesting that these residues, though widely separated in linear sequence, are close together in three dimensional space.

Prince and Sine[79] recently used γ/δ chimeras to identify residues that confer the 40-fold higher affinity of carbamoylcholine for the αδ over the αγ site. Four pairs of residues were found to contribute to selectivity, two coincident with those for conferring conotoxin M1 selectivity (δS36/γK34 and δI178/γF172), plus δD59/γE57 and δY117/γC115. δD59/γE57 contributes in a state-specific manner, preferentially affecting the desensitized state. As might be expected for agonist determinants, δS36/γK34, δD59/γE57 and δI178/γF172 are well conserved across species, but δY117/γC115 is more variable, perhaps reflecting subtle contributions to the binding site. The overall findings from studies of DMT, conotoxin M1, and carbamoylcholine selectivity determinants suggest that four linearly distant regions converge to form the (–) face of the ligand binding interface.

Studies of congenital myasthenic syndromes (see chapter 8 in this volume) also highlight the importance of the contribution of non-α subunits to the binding site. The mutation εP121L causes a congenital myasthenic syndrome by dramatically decreasing ACh affinity for the functional states, open and desensitized; in εP121L receptors, entry to these functional states occurs more slowly and with much lower probability compared to wild type.[80] P121 is invariant across all members of the cys-loop receptor superfamily, and is near residues that contribute to DMT, conotoxin M1, and carbamoylcholine selectivity. Owing to conformational restriction of its imino ring, P121 may hold the local peptide chain in a xessential for stabilizing ACh in the binding site.

Close to the carbamoylcholine/conotoxin M1 selectivity determinant, δI178/γF172, is δD180/γD174 which cross-linking studies showed to be within 9 Å of αC192-C193.[14] The mutation δD180N and the corresponding mutation γD174N decrease agonist affinity by more than 100-fold, while having much smaller effects on a series of antagonists.[81] Just C-terminal to δD180/γD174 is the conserved acidic residue, δE189/γE183. While δE189 was not cross linked to αC192-C193, the mutation δE189Q markedly decreased agonist affinity.[82] Together, δD180 and δE189 may contribute to a negative subsite within the binding cleft, stabilizing the quaternary ammonium moiety of ACh.[82]

Mutation of the conserved tryptophan at position γW55/δW57 has also been investigated. While δW57L revealed little effect, γW55L markedly increased the EC_{50} for ACh and the K_i for curare antagonism.[83] Thus, at least in the γ subunit, W55 appears to contribute to both agonist and antagonist binding.

Non-α subunit determinants of ligand affinity have also been identified in neuronal AChR. Using a series of β_4/β_2 chimeras coexpressed with α_3, Figl et al[84] and Cohen et al[85] localized determinants of relative ACh, TMA and cytisine sensitivity to the first 120 residues of the β subunits. The major contribution to ACh affinity was included within the segment β_2 104-120 (β_4 106-122), but these residues alone could not completely account for TMA and cytisine sensitivity. Using a similar approach, Wheeler et al[86] localized relative cytisine, nicotine and neuronal bungarotoxin sensitivity in $\alpha_4\beta_2$ and $\alpha_4\beta_4$ receptors to the first 80 amino acids of β_2. Recently, Harvey and Luetje[71] showed that residues 54-63 of β_2 and β_4 account for higher affinity of neuronal bungarotoxin and dihydro-β-erythroidine (DHβE) for $\alpha_3\beta_2$ compared with $\alpha_3\beta_4$ receptors. Within this segment, β_2T59 was found to be critical for high affinity antagonist binding. Minor determinants of DHβE and neuronal bungarotoxin selectivity were localized to residues 1-54, and an additional determinant of neuronal bungarotoxin sensitivity was found to lie between residues 74-80.

Subunits forming homomeric receptors potentially contain determinants equivalent to those in both α and in γ or δ. Corringer et al[87] recently tested this idea using a homomer-forming chimera containing the N-terminal extracellular domain of α_7 followed by 5HT-3 sequence. They found that mutation of α_7W54 or α_7Q56 markedly decreased ACh, nicotine and DHβE affinities, but that mutation of α_7M57 and α_7Y58 was largely without effect. These data indicate both the (+) and (−) faces of α_7 contribute to the binding site, and support the contribution of the region around γW55 and its equivalents to the agonist binding site.

A Four Loop Model for Non-α Subunit Binding Site Contributions

As observed for the α subunit, binding determinants in the non-α subunits cluster into separate regions of the linear sequence, leading to a four loop model for the contributions of the non-α subunits to the agonist binding sites[79] (Fig. 3.4). Loop 1 contains the agonist and conotoxin selectivity determinant γK34/δS36.[78,79] Loop 2 contains γW55/δW57, which was originally identified by affinity labeling in γ and δ[44,74] and subsequently by mutagenesis studies.[83,87] In addition, loop 2 contains γE57/δD59 and the analogous residues β_2T59/β_4K61 and α_7Q56 which mutagenesis studies showed to contribute to agonist and antagonist binding.[71,79,87] Loop 3 encompasses several binding site determinants. The most N-terminal of these is γS111/δY113 which contributes to α-conotoxin M1 sensitivity[78] and which is photolabeled by curare.[46] Next is γC115/δY117, which contributes to carbamoylcholine sensitivity.[79] The equivalent residue in ε, εY115, affects the assembly of the αε interface,[88] further supporting location at the (−) face of the ε subunit. Immediately adjacent are the pair of curare selectivity determinants γI116/δV118 and γY117/δT119.[76] γY117 is also photolabeled by curare.[44] The most C-terminal residue in loop 3 is P121, which contributes to agonist affinity of the desensitized and open channel states.[80] Determinants in loop 3 are also implicated in agonist sensitivity in $\alpha_3\beta_2$ and $\alpha_3\beta_4$ neuronal receptors, but the individual amino acid contributions have not been determined.[84,85] Finally, loop 4 contains the conotoxin M1 and carbamoylcholine selectivity determinant γF172/δI178[78,79] and the conserved acidic residues δD180/γD174 and δE189/γE183 identified

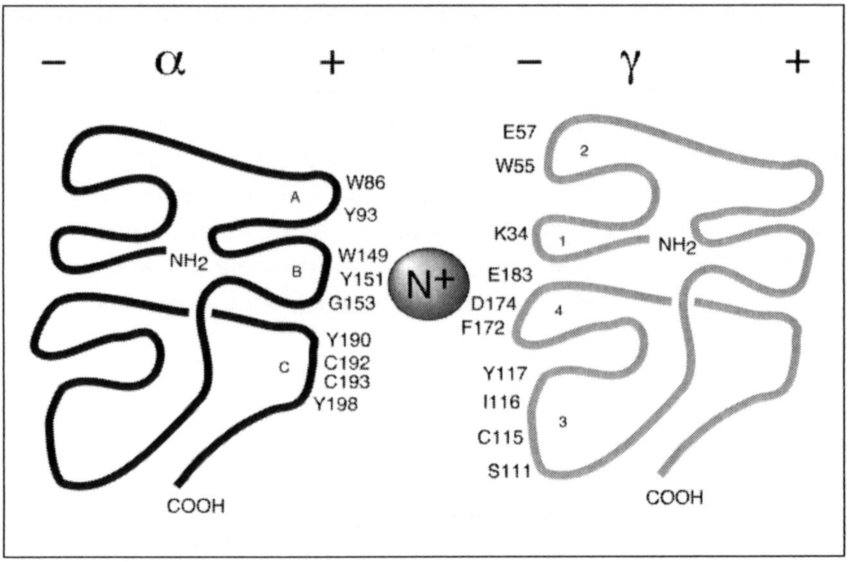

Fig. 3.4. Model of the AChR ligand binding site formed at the αγ interface of the fetal muscle receptor. Three loops from α (A-C) and four from γ (1-4) contribute to the ligand binding site. Residues labeled are those identified by affinity labeling or site directed mutagenesis studies. The second binding site in the fetal receptor is formed at the structurally homologous αδ interface. In this figure we speculatively show the agonist quaternary ammonium moiety (N^+) stabilized by conserved aromatic residues in α and acidic residues in γ. However, as detailed in the text, whether these or other residues form the primary docking site is as yet unknown.

by crosslinking and mutagenesis.[82] As discussed above, γK34/δS36 interacts strongly with γF172/δI178,[78,79] suggesting that loops 1 and 4 are closely apposed.

Summary

The weight of evidence from affinity labeling and mutagenesis studies suggests that the ligand binding sites of the AChR are located at interfaces of the α with the neighboring non-α subunits. Ligand binding sites in the various subtypes of AChR are similarly polarized, with at least three loops of the α subunit contributing to the (+) face and four loops of the non-α subunit contributing to the (−) face. Differences in agonist and antagonist sensitivity between subtypes of receptor and between binding sites within a single receptor are owing to discrete sequence differences between the subunits in these loops.

Figure 3.4 depicts a model of the AChR αγ interface, summarizing current understanding of the binding sites. We speculatively place ACh in the binding cleft such that its quaternary nitrogen is stabilized by the conserved tyrosines in the α subunit and D174 and E183 in γ. However, this model carries with it several caveats. First, mutagenesis studies do not usually distinguish between direct and allosteric contributions to ligand binding. Thus where residues have been identified solely by mutagenesis (e.g., γK34, γE183), we cannot be certain that these residues are truly at the binding site. Second, affinity labeling does not distinguish between a residue in the final docking site and one in the diffusion pathway to that site. Thus, the conserved aromatic residues in the α subunit may line the aromatic gorge leading to the final docking site for ACh.

To conclude, the model in Figure 3.4 represents a starting point for future studies of AChR binding site structure. With the rapid pace of structure-function studies on the AChR, it is likely that the model will undergo considerable addition and revision. However, the ultimate test of this or any future model of the binding sites will be against a three dimensional image of the receptor which defines the positions of individual amino acids.

Acknowledgments

We would like to thank the NIH (NS 31744 to SMS) and the Myasthenia Gravis Foundation (RP) for their support.

References

1. Langley JN, Anderson HK. The actions of nicotine on the ciliary ganglion of the third cranial nerve. J Physiol (Lond) 1892; 13:460-468.
2. Dale HH. The action of certain esters and ethers of choline and their relation to muscarine. J Pharmacol Exp Ther 1914; 6:147-190.
3. Weill CL, McNamee MG, Karlin A. Affinity-labeling of purified acetylcholine receptor from *Torpedo californica.* Biochem Biophys Res Comm 1974; 61:997-1003.
4. Reynolds JA, Karlin A. Molecular weight in detergent solution of acetylcholine receptor from *Torpedo californica.* Biochemistry 1978; 17:2035-2038.
5. Raftery MA, Hunkapiller MW, Strader CD et al. Acetylcholine receptor: Complex of homologous subunits. Science 1980; 208: 1454-1456.
6. Huganir RL, Schell MA, Racker E. Reconstitution of the purified acetylcholine receptor from *Torpedo californica.* FEBS Lett 1979; 108:155-160.

7. Nelson N, Anholt R, Lindstrom J et al. Reconstitution of purified acetylcholine receptors with functional ion channels in planar lipid bilayers. Proc Natl Acad Sci USA 1980; 77:3057-3061.
8. Holtzman E, Wise D, Wall J et al. Electron microscopy of complexes of isolated acetylcholine receptor, biotinyl-toxin, and avidin. Proc Natl Acad Sci U S A 1982; 79:310-314.
9. Karlin A, Holtzman E, Yodh N et al. The arrangement of the subunits of the acetylcholine receptor of *Torpedo californica.* J Biol Chem 1983; 258:6678-81.
10. Kurosaki T, Fukuda K, Konno T et al. Functional properties of nicotinic acetylcholine receptor subunits expressed in various combinations. FEBS Lett 1987; 214:253-258.
11. Blount P, Merlie JP. Molecular basis of the two nonequivalent ligand binding sites of the muscle nicotinic acetylcholine receptor. Neuron 1989; 3:349-357.
12. Sine SM, Claudio T. Gamma- and delta-subunits regulate the affinity and the cooperativity of ligand binding to the acetylcholine receptor. J Biol Chem 1991; 266:19369-19377.
13. Pedersen SE, Cohen JB. d-Tubocurarine binding sites are located at alpha-gamma and alpha-delta subunit interfaces of the nicotinic acetylcholine receptor. Proc Natl Acad Sci U S A 1990; 87:2785-2789.
14. Czajkowski C, Karlin A. Structure of the nicotinic receptor acetylcholine-binding site. Identification of acidic residues in the delta subunit within 0.9 nm of the alpha subunit binding site disulfide. J Biol Chem 1995; 270:3160-3164.
15. Kubalek E, Ralston S, Lindstrom J et al. Location of subunits within the acetylcholine receptor by electron image analysis of tubular crystals from *Torpedo marmorata.* J Cell Biol 1987; 105:9-18.
16. Karlin A, Akabas MH. Toward a structural basis for the function of nicotinic acetylcholine receptors and their cousins. Neuron 1995; 15:1231-1244.
17. Hucho F, Tsetlin VI, Machold J. The emerging three-dimensional structure of a receptor. The nicotinic acetylcholine receptor. Eur J Biochem 1996; 239:539-557.
18. Criado M, Sarin V, Fox JL et al. Evidence that the acetylcholine binding site is not formed by the sequence alpha 127-143 of the acetylcholine receptor. Biochemistry 1986; 25:2839-4286.
19. Sumikawa K, Gehle VM. Assembly of mutant subunits of the nicotinic acetylcholine receptor lacking the conserved disulfide loop structure. J Biol Chem 1992; 267:6286-6290.
20. Fu DX, Sine SM. Asymmetric contribution of the conserved disulfide loop to subunit oligomerization and assembly of the nicotinic acetylcholine receptor. J Biol Chem 1996; 271:31479-31484.
21. Karlin A. Structure of nicotinic acetylcholine receptors. Curr Opin Neurobiol 1993; 3:299-309.
22. Noda M, Takahashi H, Tanabe T et al. Primary structure of alpha-subunit precursor of *Torpedo californica* acetylcholine receptor deduced from cDNA sequence. Nature 1982; 299:793-797.

23. Noda M, Takahashi H, Tanabe T et al. Structural homology of *Torpedo californica* acetylcholine receptor subunits. Nature 1983; 302:528-32.
24. Noda M, Takahashi H, Tanabe T et al. Primary structures of beta- and delta-subunit precursors of *Torpedo californica* acetylcholine receptor deduced from cDNA sequences. Nature 1983; 301:251-255.
25. Ballivet M, Patrick J, Lee J et al. Molecular cloning of cDNA coding for the gamma subunit of Torpedo acetylcholine receptor. Proc Natl Acad Sci U S A 1982; 79:4466-4470.
26. Takai T, Noda M, Mishina M et al. Cloning, sequencing and expression of cDNA for a novel subunit of acetylcholine receptor from calf muscle. Nature 1985; 315:761-764.
27. Le Novere N, Changeux JP. Molecular evolution of the nicotinic acetylcholine receptor: An example of a multigene family in excitable cells. J Mol Evol 1995; 40:155-172.
28. Elgoyhen AB, Johnson DS, Boulter J et al. Alpha 9: An acetylcholine receptor with novel pharmacological properties expressed in rat cochlear hair cells. Cell 1994; 79:705-715.
29. Ravdin P, Berg DK. Inhibition of neuronal acetylcholine sensitivity by α-neurotoxins from *Bungarus multicinctus* venom. Proc Natl Acad Sci USA 1979; 76:2072-2076.
30. Taylor P. Agents acting at the neuromuscular junction and autonomic ganglia. In: Hardman JG, Limbird LE, eds. Goodman and Gilman's The Pharmacological Basis of Therapeutics (9th Edition). New York: McGraw-Hill, 1996:177-197.
31. Sullivan JP, Decker MW, Brioni JD et al. (±)-Epibatidine elicits a diversity of in vitro and in vivo effects mediated by nicotinic acetylcholine receptors. J Pharmacol Exp Ther 1994; 271:624-631.
32. Gerzanich V, Peng X, Wang F et al. Comparative pharmacology of epibatidine: A potent agonist for neuronal nicotinic receptors. Mol Pharmacol 1995; 48:774-782.
33. Beers WH, Reich E. Structure and activity of acetylcholine. Nature 1970; 228:917-922.
34. Sheridan RP, Nilakantan R, Dixon JS et al. The ensemble approach to distance geometry: Application to the nicotinic pharmacophore. J Med Chem 1986; 29:899-906.
35. Cockcroft VB, Osguthorpe DJ, Barnard EA et al. Modeling of agonist binding to the ligand-gated ion channel superfamily of receptors. Proteins 1990; 8:386-397.
36. Jackson MB. Perfection of a synaptic receptor: Kinetics and energetics of the acetylcholine receptor. Proc Natl Acad Sci USA 1989; 86:2199-2203.
37. Lingle CJ, Maconochie D, Steinbach JH. Activation of skeletal muscle nicotinic acetylcholine receptors. J Memb Biol 1992; 126:195-217.
38. Galzi JL, Revah F, Bessis A et al. Functional architecture of the nicotinic acetylcholine receptor: From electric organ to brain. Ann Rev Pharmacol Toxicol 1991; 31:37-72.

39. Galzi JL, Revah F, Black D et al. Identification of a novel amino acid alpha-tyrosine 93 within the cholinergic ligand-binding sites of the acetylcholine receptor by photoaffinity labeling. Additional evidence for a three-loop model of the cholinergic ligand-binding sites. J Biol Chem 1990; 265:10430-10437.
40. Cohen JB, Sharp SD, Liu WS. Structure of the agonist-binding site of the nicotinic acetylcholine receptor. [^{3}H]acetylcholine mustard identifies residues in the cation-binding subsite. J Biol Chem 1991; 266:23354-23364.
41. Dennis M, Giraudat J, Kotzyba-Hibert F et al. Amino acids of the *Torpedo marmorata* acetylcholine receptor alpha subunit labeled by a photoaffinity ligand for the acetylcholine binding site. Biochemistry 1988; 27:2346-2357.
42. Abramson SN, Li Y, Culver P et al. An analog of lophotoxin reacts covalently with Tyr190 in the alpha-subunit of the nicotinic acetylcholine receptor. J Biol Chem 1989; 264:12666-12672.
43. Middleton RE, Cohen JB. Mapping of the acetylcholine binding site of the nicotinic acetylcholine receptor: [^{3}H]nicotine as an agonist photoaffinity label. Biochemistry 1991; 30:6987-6997.
44. Chiara DC. Structural studies of the nicotinic acetylcholine receptor, a ligand gated ion channel. PhD Thesis Washington University Missouri 1996.
45. Mosckovitz R, Gershoni JM. Three possible disulfides in the acetylcholine receptor alpha-subunit. J Biol Chem 1988; 263:1017-1022.
46. Kao PN, Dwork AJ, Kaldany RR et al. Identification of the alpha subunit half-cystine specifically labeled by an affinity reagent for the acetylcholine receptor binding site. J Biol Chem 1984; 259: 11662-11665.
47. Ramirez-Latorre J, Yu CR, Qu X et al. Functional contributions of alpha 5 subunit to neuronal acetylcholine receptor channels. Nature 1996; 380:347-351.
48. Wang F, Gerzanich V, Wells GB et al. Assembly of human neuronal nicotinic receptor alpha5 subunits with alpha3, beta2, and beta4 subunits. J Biol Chem 1996; 271:17656-17665.
49. Kreienkamp HJ, Sine SM, Maeda RK et al. Glycosylation sites selectively interfere with alph- binding to the nicotinic acetylcholine receptor. J Biol Chem 1994; 269:8108-8114.
50. Sine SM, Ohno K, Bouzat C et al. Mutation of the acetylcholine receptor alpha subunit causes a slow-channel myasthenic syndrome by enhancing agonist binding affinity. Neuron 1995; 15:229-239.
51. Mishina M, Tobimatsu T, Imoto K et al. Location of functional regions of acetylcholine receptor alpha-subunit by site-directed mutagenesis. Nature 1985; 313:364-369.
52. Tomaselli GF, McLaughlin JT, Jurman ME et al. Mutations affecting agonist sensitivity of the nicotinic acetylcholine receptor. Biophys J 1991; 60:721-727.

53. Galzi JL, Bertrand D, Devillers-Thiery A et al. Functional significance of aromatic amino acids from three peptide loops of the alpha 7 neuronal nicotinic receptor site investigated by site-directed mutagenesis. FEBS Lett 1991; 294:198-202.
54. Sine SM, Quiram P, Papanikolaou F et al. Conserved tyrosines in the alpha subunit of the nicotinic acetylcholine receptor stabilize quaternary ammonium groups of agonists and curariform antagonists. J Biol Chem 1994; 269:8808-8816.
55. Nowak MW, Kearney PC, Sampson JR et al. Nicotinic receptor binding site probed with unnatural amino acid incorporation in intact cells. Science 1995; 268:439-442.
56. Filatov GN, Aylwin ML, White MM. Selective enhancement of the interaction of curare with the nicotinic acetylcholine receptor. Mol Pharmacol 1993; 44:237-241.
57. Aylwin ML, White MM. Gating properties of mutant acetylcholine receptors. Mol Pharmacol 1994; 46:1149-1155.
58. Chen J, Zhang Y, Akk G et al. Activation kinetics of recombinant mouse nicotinic acetylcholine receptors: Mutations of alpha-subunit tyrosine 190 affect both binding and gating. Biophys J 1995; 69:849-859.
59. Unwin N. Nicotinic acetylcholine receptor at 9 Å resolution. J Mol Biol 1993; 229:1101-1124.
60. Sussman JL, Harel M, Frolow et al. Atomic structure of acetylcholinesterase from *Torpedo californica*: A prototypic acetylcholine-binding protein. Science 1991; 253:872-879.
61. Damle VN, McLaughlin M, Karlin A. Bromoacetylcholine as an affinity label of the acetylcholine receptor from *Torpedo californica*. Biochem Biophys Res Comm 1978; 84:845-851.
62. Damle VN, Karlin A. Affinity labeling of one of two alpha-neurotoxin binding sites in acetylcholine receptor from *Torpedo californica*. Biochemistry 1978; 17:2039-2045.
63. Neubig RR, Cohen JB. Equilibrium binding of [^{3}H]tubocurarine and [^{3}H]acetylcholine by Torpedo postsynaptic membranes. Biochemistry 1979; 18:5464-5475.
64. Sine SM, Taylor P. Relationships between reversible antagonist occupancy and the functional capacity of the acetylcholine receptor. J Biol Chem 1981; 256:6692-6699.
65. Sine SM, Claudio T, Sigworth FJ. Activation of Torpedo acetylcholine receptors expressed in mouse fibroblasts. Single channel current kinetics reveal distinct agonist binding affinities. J Gen Physiol 1990; 96:395-437.
66. Zhang Y, Chen J, Auerbach A. Activation of recombinant mouse acetylcholine receptors by acetylcholine, carbamoylcholine and tetramethylammonium. J Physiol 1995; 486:189-206.
67. Prince RJ, Sine SM. Epibatidine selects between the binding sites of muscle nicotinic acetylcholine receptors. Soc Neurosci Abs 1996; 22:110.3.

68. Blount P, Merlie JP. Native folding of an acetylcholine receptor alpha subunit expressed in the absence of other receptor subunits. J Biol Chem 1988; 263:1072-80.
69. Luetje CW, Wada K, Rogers S et al. Neurotoxins distinguish between different neuronal nicotinic acetylcholine receptor subunit combinations. J Neurochem 1990; 55:632-640.
70. Luetje CW, Patrick J. Both alpha- and beta-subunits contribute to the agonist sensitivity of neuronal nicotinic acetylcholine receptors. J Neurosci 1991; 11:837-845.
71. Harvey SC, Luetje CW. Determinants of competitive antagonist sensitivity on neuronal nicotinic receptor beta subunits. J Neurosci 1996; 16:3798-3806.
72. Oswald RE, Changeux JP. Crosslinking of alpha-bungarotoxin to the acetylcholine receptor from *Torpedo marmorata* by ultraviolet light irradiation. FEBS Lett 1982; 139:225-229.
73. Langenbuch-Cachat J, Bon C, Mulle C et al. Photoaffinity labeling of the acetylcholine binding sites on the nicotinic receptor by an aryldiazonium derivative. Biochemistry 1988; 27:2337-2345.
74. Chiara DC, Cohen JB. Identification of amino acids contributing to the high and low affinity d-tubocurarine (dTC) sites on the Torpedo nicotinic acetylcholine receptor (nAChR) subunits. Biophys J 1992; 61:A106.
75. Czajkowski C, Karlin A. Agonist binding site of Torpedo electric tissue nicotinic acetylcholine receptor. A negatively charged region of the delta subunit within 0.9 nm of the alpha subunit binding site disulfide. J Biol Chem 1991; 266:22603-22612.
76. Sine SM. Molecular dissection of subunit interfaces in the acetylcholine receptor: Identification of residues that determine curare selectivity. Proc Natl Acad Sci USA 1993; 90:9436-9440.
77. Fu DX, Sine SM. Competitive antagonists bridge the alpha-gamma subunit interface of the acetylcholine receptor through quaternary ammonium-aromatic interactions. J Biol Chem 1994; 269:26152-26157.
78. Sine SM, Kreienkamp HJ, Bren N et al. Molecular dissection of subunit interfaces in the acetylcholine receptor: Identification of determinants of alpha-conotoxin M1 selectivity. Neuron 1995; 15:205-211.
79. Prince RJ, Sine SM. Molecular dissection of subunit interfaces in the acetylcholine receptor: Identification of residues that determine agonist selectivity. J Biol Chem 1996; 271:25770-25777.
80. Ohno K, Wang H-L, Milone M et al. Congenital myasthenic syndrome caused by decreased agonist binding affinity due to a mutation in the acetylcholine receptor ε subunit. Neuron 1996; 17:157-170.
81. Martin M, Czajkowski C, Karlin A. The contributions of aspartyl residues in the acetylcholine receptor gamma and delta subunits to the binding of agonists and competitive antagonists. J Biol Chem 1996; 271:13497-13503.

82. Czajkowski C, Kaufmann C, Karlin A. Negatively charged amino acid residues in the nicotinic receptor δ subunit that contribute to the binding of acetylcholine. Proc Natl Acad Sci USA 1993; 90:6285-6289.
83. O'Leary ME, Filatov GN, White MN. Characterization of d-tubocurarine binding site of Torpedo acetylcholine receptor. Am J Physiol 1994; 266:C648-653.
84. Figl A, Cohen BN, Quick MW et al. Regions of beta 4.beta 2 subunit chimeras that contribute to the agonist selectivity of neuronal nicotinic receptors. FEBS Lett 1992; 308:245-248.
85. Cohen BN, Figl A, Quick MW et al. Regions of beta 2 and beta 4 responsible for differences between the steady state dose-response relationships of the alpha 3 beta 2 and alpha 3 beta 4 neuronal nicotinic receptors. J Gen Physiol 1995; 105:745-764.
86. Wheeler SV, Chad JE, Foreman R. Residues 1 to 80 of the N-terminal domain of the beta subunit confer neuronal bungarotoxin sensitivity and agonist selectivity on neuronal nicotinic receptors. FEBS Lett 1993; 332:139-142.
87. Corringer PJ, Galzi JL, Eisele JL et al. Identification of a new component of the agonist binding site of the nicotinic alpha 7 homo-oligomeric receptor. J Biol Chem 1995; 270:11749-11752.
88. Gu Y, Camacho P, Gardner P et al. Identification of two amino acid residues in the epsilon subunit that promote mammalian muscle acetylcholine receptor assembly in COS cells. Neuron 1991; 6:879-887.

CHAPTER 4

Adding up the Energies in the Acetylcholine Receptor Channel: Relevance to Allosteric Theory

Meyer B. Jackson

The acetylcholine receptor (AChR) is a protein that spans the cell membrane to form an aqueous pore through which ions flow. The pore or channel is normally closed, but opens when agonists bind to stereospecific binding sites on the protein surface. Ligand binding also desensitizes receptors by converting them to a conformation in which the channel is once again closed. Thus, by controlling the conformational state of this protein, the neurotransmitter acetylcholine (ACh) can regulate the ionic conductance of the postsynaptic cell membrane. In this way, AChRs and other closely related proteins belonging to the same superfamily of ligand-gated channels (see chapter 2 in this volume) serve in the transduction of chemical signals to electrical signals.

How does ACh binding influence the functional state of the channel? This is the central question in allosteric theory, and the basic relationship between conformational state and ligand binding places ligand-gated channels in the realm of allosteric proteins. Theories for allosteric proteins were originally developed in order to understand cooperative interactions in multisubunit enzymes, years before the molecular nature of ligand-gated channels was appreciated. However, ligand-gated channels have emerged as ideal subjects for the study of molecular signaling mechanisms, thanks in part to the patch-clamp technique, which provides measurements of the conformational state of single protein molecules. This has made the

The Nicotinic Acetylcholine Receptor: Current Views and Future Trends,
edited by Francisco J. Barrantes.

allosteric formalism a powerful conceptual tool in the study of signal transduction at cell membranes. This chapter will examine the assumptions underlying allosteric theory and interpret these assumptions in terms of specific molecular contacts between ligands and specific residues of the protein. One of the key assumptions of allosteric theory is that specific terms in the free energy of the protein and protein-ligand complex can be added together according to a simple linear rule. This chapter will explore the possibility of basing this assumption on a more fundamental assumption that the energies of specific atomic contacts within the protein can be added up in the same way. These ideas will be illustrated by using examples from the literature on the AChR.

The Assumptions of Allosteric Theory

As originally formulated, allosteric theory made two key assumptions:[25]

1) The protein isomerizes between two stable conformations. These two conformations are distinguished both by differences in the biological function of the protein and by differences in the affinity for ligand. Both function and binding affinity change concomitantly during this transition, so that the protein is not permitted to have the functional properties of one state and the binding affinity of the other. This isomerization of the protein is called the allosteric transition, and, in principle, this transition can take place under any state of receptor occupancy.

2) In multisubunit proteins, the conformational transition is strictly cooperative; all of the subunits are in the same state. This means that the binding site of one subunit will not sense the state of occupancy of another binding site in another subunit as long as the protein does not undergo an allosteric transition. Within a given conformation, the binding sites do not interact, so that one binding site can be influenced by another only through an effect on the equilibrium between the high-affinity and low-affinity conformations.

Additional assumptions made in the classical formulation of this theory regarding symmetry and the equivalence of subunits are not essential, and can even be considered overly restrictive in light of the pseudosymmetry of homologous but nonidentical subunits found in the AChR as well as in most other ligand-gated channels. In fact, only assumption #1 is truly essential to the theory, and this assumption alone can be used to develop a simple and highly instructive model for a receptor with a single binding site.[2] With the low-affinity and high-affinity conformations linked by an allosteric transition, we have the following scheme:

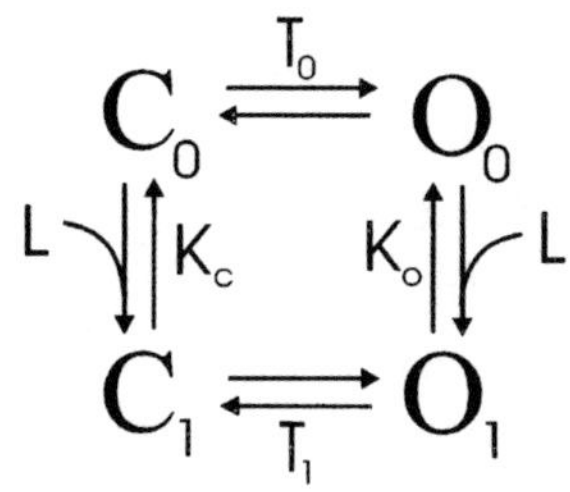

where C and O denote the closed state and open state, respectively, with subscripts indicating whether the single binding site is empty or occupied. The dissociation constant is K_c or K_O, depending on whether the channel is closed or open. The equilibrium constant for the allosteric transition is T_0 when the binding site is empty and T_1 when the binding site is occupied. Detailed balance requires that these four parameters are related as $K_c/T_1 = K_O/T_0$. The dose-response relation for a receptor described by this scheme is then:

$$R = \frac{K_c T_0 + L\,T_1}{K_c(1+T_0) + L(1+T_1)} \qquad (1)$$

This expression can be simplified by noting that T_0 is small compared to one and that T_1 is large compared to one. These approximations together with the detailed balance constraint lead to the following expression:

$$R = \frac{L R_{max}}{L + \frac{K_o}{T_0}} \qquad (2)$$

where $R_{max} = T_1/(1 + T_1)$ is the maximal response. The concentration of agonist that will produce a half-maximal response is K_O/T_0 and this is taken as an apparent affinity of the receptor. The instructive value of this one-binding-site model is that the dose-response relation shows how the equilibrium constant of the allosteric transition can influence the sensitivity of the receptor. Anything that perturbs this equilibrium will change T_0 and shift the dose-response curve. Thus, the apparent affinity of the receptor can change even as K_O, a true binding constant, remains constant. Monod et al[1] also made the point that changes in the equilibrium constant of the allosteric transition can manifest as an alteration in sensitivity. However, at the time that this theory was published, there was little experimental evidence that this could happen. Now we have many good examples, some of which will be discussed below.

The basic idea of an allosteric receptor is illustrated for a ligand-gated channel with two binding sites (Fig. 4.1). The two conformations of the protein differ in two ways. In terms of functional state, the channel is either open or closed, and the isomerization of the unligated receptor in Figure 4.1A depicts the allosteric transition. Furthermore, the binding sites change during the transition between the two conformations, such that ligand binds weakly when the channel is closed and tightly when the channel is open. The tighter binding of agonist to the open-channel conformation drives the transition. This is the essence of allosteric theory.

The first assumption of allosteric theory is supported by patch-clamp studies of the AChR.[3,4] The allosteric transition between the open and closed states is readily seen as spontaneous openings in the absence of ligand. Very brief closures known as flickers reflect the same transition in the fully ligated receptor.[5] The desensitized state can also be detected as a small fraction of high-affinity receptor present prior to the presentation of ligand.[6] Evidence from a variety of sources revealed that the structure of the binding site changes during the gating transition.[7] Changes in binding site structure have been seen in electron microscope images of the receptor.[8] Comparisons of the structures of tethered agonists and antagonists suggest that channel opening is accompanied by a shortening of the distance between hydrogen bonding and charged regions of the receptor binding site.[9] The chemical reactivity of residues at the binding site can also be altered by agonists.[10]

Testing the second assumption has been more difficult than testing the first assumption. Does occupation of one site of a ligand-gated channel influence the other binding site in the absence of a conformational transition? This question becomes more complicated when one considers that the closed conformation of the protein binds two agonist molecules with affinities differing by a factor of more than 100.[11,12] This could be due to intrinsic structural differences between the two binding sites, as analyzed in detail in chapter 3 of this volume, which would not violate the assumptions above. However, the different affinities could also reflect anticooperative interactions

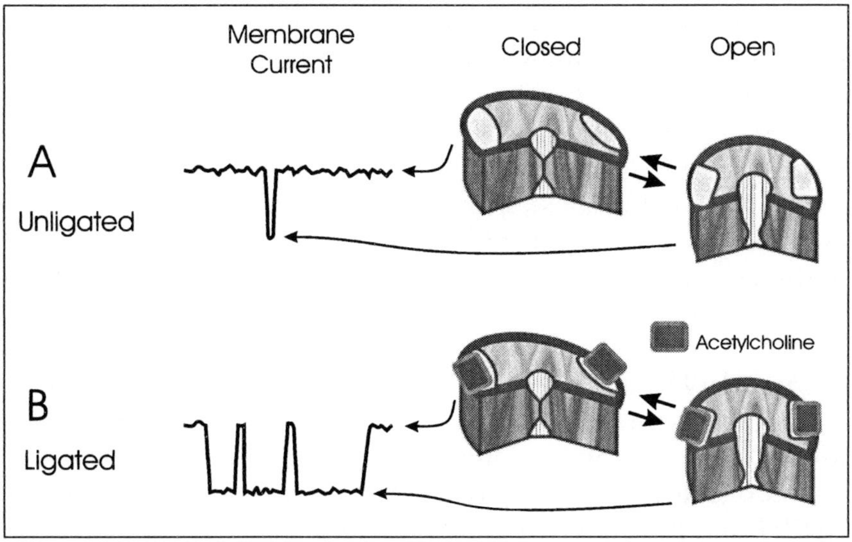

Fig. 4.1. The activation of the AChR is depicted as a conformational transition of an allosteric protein. The opening of the channel produces a negative-going step in membrane current; the dashed arrows indicate the current level of the corresponding state of the channel. (A) In the absence of ligand the channel is closed most of the time and exhibits occasional brief openings. (B) When the binding sites are occupied, the channel is open most of the time and exhibits occasional brief closures. Note that the open conformation of the binding site is depicted as providing a better fit for acetylcholine. This improvement in binding allows the ligand to drive the conformational transition.

between the two binding sites in the absence of the allosteric transition, and this would be a direct contradiction of the second assumption above. The first possibility of intrinsic differences has some support from experiments. Although each binding site of the AChR contains residues from one of the two α subunits, different subunits flank the α subunits; residues from the γ subunit contribute to one binding site and residues from the δ subunit contribute to the other (see chapter 3 in this volume and refs. 7, 13). Electron microscope images have revealed differences between the structures of the two putative agonist binding domains of the channel in its closed state,[8,14] and this also supports the idea that the two binding sites have different affinities prior to the binding of an agonist. The difference between the affinities of the two binding sites therefore need not imply negative interactions within the closed conformation. Thus, the second assumption of allosteric theory is consistent with the available data, but a more stringent test of binding site interactions would be welcome.

In summary, the existence of an allosteric transition is well supported by experiments, and this is a key element of allosteric theory. It is likely that the allosteric transition serves as the principal vehicle for interactions between the binding sites, but the possibility of direct interactions between binding sites is difficult to rule out entirely at present.

Allosteric Theory in Terms of Macromolecular Additivity

We can gain some insight into the assumptions underlying allosteric theory by thinking about the free energy changes during the transitions depicted in Figure 4.1. The allosteric transition has an equilibrium constant of T_o, and therefore the intrinsic free energy difference between the open and closed conformations of the protein is given by $\Delta G_\iota = -RT \ln T_0$. This free energy difference determines which conformation is most abundant in the absence of ligand. Clearly, that should be the closed state of the channel, which will make ΔG_ι positive.

The free energy difference between the open and closed states of the ligated protein (Fig. 4.1B) can then be thought of as a sum of two terms. One of these terms would be ΔG_ι for the conformational change in the absence of ligand. The other free energy term is the change in binding energy associated with the conformational transition, ΔG_λ. This term is actually a $\Delta\Delta G$ because it is the difference between the ΔG of binding to the open state and the ΔG of binding to the closed state; i.e., $\Delta G_\lambda = \Delta G_{b\text{-}o} - \Delta G_{b\text{-}c}$. In the absence of ligand, a positive value of ΔG_ι keeps the channel closed most of the time, and all that is seen in a patch-clamp experiment is occasional spontaneous openings (Fig. 4.1A). For the occupation of *n* binding sites by agonist molecules, allosteric theory dictates that the total change in free energy associated with the conformational transition will be a simple sum of the intrinsic term and the binding term,

$$\Delta G = \Delta G_\iota + n\Delta G_\lambda \qquad (3)$$

Returning to assumption #2 above, we can see that if occupation of one binding site changes the properties of other sites in the absence of the allosteric transition, then ΔG_λ for a particular site would vary and the additivity of free energies would break down.

Thus, the assumption of additivity of the free energies, ΔG_ι and ΔG_λ, as expressed in Equation 3, can be viewed as an alternative formulation of the second assumption of allosteric theory stated above. We will refer to this assumption as macromolecular additivity and keep it in mind as we move on to extend the assumption of additivity to a microscopic level below.

Following up on the logic of macromolecular additivity leads to a straightforward thermodynamic explanation for the ligand-induced conformational change of an allosteric protein. If the open state binds ligands more tightly than the closed state, then ΔG_λ will be negative. If $-n\Delta G_\lambda > \Delta G_\iota$ then $\Delta G_\iota + n\Delta G_\lambda$ will be negative and the channel will open when n binding sites are occupied by agonist. The ligand binding energy must then be sufficient to overcome the intrinsic free energy that opposes channel opening. Indeed, estimates of the maximum energy that can be derived from binding acetylcholine have suggested that binding only one molecule cannot provide enough energy to gate the channel, and that equipping the AChR with two binding sites may be necessary to overcome the energetic barrier to opening.[15]

Contacts in Allosteric Proteins: Micromolecular Additivity

Since the open and closed channel conformations of ligand-gated channels are stable on time scales long compared to the picosecond time scales of stochastic motions in proteins, there must be stable noncovalent interactions that maintain each conformation. During a conformational transition, one set of internal contacts within the protein breaks as another set of contacts forms. These contacts are thus relevant to the intrinsic ΔG_ι of the conformational change in the absence of ligand, just as contacts between the ligand and the binding site are relevant to ΔG_λ. Considering the assumption of additivity of these two free energy terms for the whole protein, we might wonder whether the energies of the individual contacts that comprise these free energies obey a similar rule of additivity. Then there would be no cooperative interactions between separate contacts and ΔG_ι could be expressed as a sum of the form

$$\Delta G_\iota = \Sigma g_{io} - \Sigma g_{ic} \qquad (4)$$

where a contact in the closed state makes a contribution g_{ic} to the stability of the closed state, and a contact in the open state makes a contribution g_{io} to the stability of the open state. Likewise, contacts between the ligand and the closed state or open state can be denoted by the subscripts bc and bo, respectively. The binding energies can then be represented by the expressions:

$$\Delta G_{bc} = \Sigma g_{bc} \tag{5a}$$
$$\Delta G_{bo} = \Sigma g_{bo} \tag{5b}$$

and ΔG_λ becomes

$$\Delta G_\lambda = \Sigma g_{bo} - \Sigma g_{bc} \tag{6}$$

We will refer to the assumption that energies of individual contacts can be added together in this manner as micromolecular additivity, to draw on an analogy with macromolecular additivity as embodied in Equation 3 above. Clearly, if each of these gs are independent additive terms, then ΔG_ι and ΔG_λ can be added as in the previous section. Thus, micromolecular additivity implies macromolecular additivity. However, violations of micromolecular additivity need not negate macromolecular additivity, and this point will be discussed further below.

A strategy for understanding the energy balance between ΔG_ι and ΔG_λ can then be developed in terms of micromolecular additivity by identifying the relevant contacts, determining, calculating, or guessing their energies, and then adding up the terms. In this way the assumption of micromolecular additivity not only provides a conceptual framework for allosteric theory, but also aids in interpreting changes in the functional properties of an allosteric protein resulting from a mutation or chemical modification. The decomposition of free energy changes into a sum of terms in Equations 4, 5 and 6 may be useful in interpreting changes in relative stability. Thus, a change in protein structure achieved through molecular genetic manipulation, as analyzed in chapter 3 of this volume, may alter contacts that participate in ligand binding, or conformational stability, or both, and measurements of the relevant free energy changes can be interpreted with the aid of these equations.

These formulas form a useful framework for analyzing changes in equilibrium constants. However, they offer no help in interpreting changes in rate. Of course, one can expect that strengthening contacts that stabilize a conformation will reduce the rate of exit from that conformation, but the quantitative relationship is not obvious, even when one makes the simplifying assumption of additivity. To understand the kinetics of conformational transitions in proteins we can consider a model in which the transition is obstructed by a specified number of independent groups, each of which can isomerize independently into a configuration which no longer obstructs the transition.[16] Like the tumblers of a lock, only when all of these groups are out of the way can the transition take place. The movement of these obstructions was originally proposed to be rotation of bonds, but within the framework of a stochastic process, breaking of contacts can be treated in the same way, as long as they are independent. For N identical obstacles to the transition with a rate, α, of flipping out of the way, and a rate, β, of returning to block the transition (such that the energy between the two states of the obstacle is $E = -RT \ln(\alpha/\beta)$), the probability that all of the obstacles are out of the way is

$$P_n = \frac{1}{(1+e^{E/RT})^N} \tag{7}$$

The rate of the conformational transition will then be equal to this probability times the intrinsic rate of the unobstructed transition, ϕ

$$r = \frac{\phi}{(1+e^{E/RT})^N} \tag{8}$$

The intrinsic rate, ϕ, for such a transition in proteins can be taken as similar to the rate of rotational isomerizations in polymers and is thought to be on the order of 10^9 sec^{-1}.[17] Equation 8 differs from an earlier expression for the rate of such a process[16] because of the manner in which the problem was set up, but the essence of the model is unchanged and the present form has the advantage of being simpler and easier to extend to nonidentical contacts, as will be illustrated below.

With nanosecond time scales for the dynamics of individual contacts and obstacles, one can obtain millisecond time scales for the rate of the conformational transition when $E = 2.3RT$ and $N = 6$ or 7. When $E = 0$, millisecond time scales can be achieved with N of about 26. Thus, for this model it is easy to find parameter values that give transition rates typical of those observed in the gating of ion channels. Furthermore, even though the model has a formal solution that is a sum of N exponentials, for choices of N and E that produce millisecond time scales for the conformational transition, the time course of the transition is dominated by a single slow exponential with a decay constant very close to that given by Equation 8.[16] Studies of single-channel kinetics have established exponential time courses for the gating transitions. Thus, this model can account for the qualitative features of the kinetics of ligand-gated channels.

It is easy to generalize the tumbler lock model to a system with nonequivalent obstacles, and this may make the model more relevant to proteins in which the obstacles to a conformational transition include both bonds that must rotate out of the way and conformation stabilizing contacts that must break. The form of Equation 8 readily suggests the following generalization to nonequivalent obstacles

$$r = \phi \prod_{i}^{N} \frac{1}{1 + e^{E_i/RT}} \qquad (9)$$

Thus, if a mutation alters the energy of a particular contact from E to E', and that contact contributes to the stability of a conformation, then the rate of a transition out of this conformation would be increased by a factor of

$$r'/r = (1+e^{E/RT})/(1+e^{E'/RT}) \qquad (10)$$

Removing the contact completely would simply multiply the rate by the factor representing that contact

$$r'/r = 1+e^{E/RT} \qquad (11)$$

Equations 9, 10, and 11 provide a way to use the framework of additivity to interpret changes in rates caused by structural manipulations. These equations thus complement Equations 4-6 for interpreting changes in free energy.

Are Interactions Additive?

The assumption of micromolecular additivity of contact energies has two significant advantages. It provides a simple rationale for the assumption of macromolecular additivity, from which allosteric theory can be derived. Furthermore, micromolecular additivity leads to very simple theoretical expressions for how alterations in chemical structure change the free energies and rates of conformational transitions. These expressions could be useful in interpreting experimental results, especially in studies where amino acid replacement might be thought to alter one or a few contacts selectively, as reviewed in chapter 3 of this volume. It is therefore important to evaluate the likelihood that this assumption is valid. Among the interactions that stabilize the conformation of a protein are hydrophobic forces, electrostatic forces, hydrogen bonds, and strain and torsional potentials of covalent bonds. Based on what is known about the physical nature of these interactions, what can we say about the assumption of micromolecular additivity?

Hydrophobic Forces

When nonpolar or hydrophobic groups interact, the energy is thought to take the form of an induced dipole-induced dipole attraction. Van der Waals forces of essentially the same form hold nonpolar liquids together. The importance of three-body forces relative to pairwise interactions has been tested in theories of liquid argon.[18] Including the three-body term does improve the accuracy of calculations of thermodynamic quantities, but the error due to neglecting this term is only on the order of 10% at physiological temperatures. Furthermore, this error can be reduced substantially by using an "effective pair-potential" that corrects for the neglect of the three-body term. The energy of a hydrophobic interaction is often treated as proportional to the surface area of contact,[19] and this is in essence an additive or pairwise representation which is thought to work well in predicting tendencies of different amino acids to partition into membranes (hydropathy analysis) and to fold up into different kinds of secondary structure. Even with many-body energy terms, the $1/r^6$ dependence on distance would limit all interactions to very short distances, and this would leave all but the most proximal contacts highly additive. So for hydrophobic forces additivity appears to be a safe assumption.

Electrostatic Forces

Classical electrostatics obeys strict additivity for interaction energies. However, over short molecular distances adherence to classical electrostatics cannot be guaranteed. Furthermore, fields close to charged groups can be strong enough to saturate a dielectric. This could lead to nonadditivity over relatively short distances, but over longer distances additivity is still safe.

Hydrogen Bonds

The hydrogen bond is often treated as an electrostatic interaction. However, the electrostatic picture ignores the covalent nature of hydrogen bonds, and recent natural bond orbital calculations have suggested that the covalent properties of the hydrogen bond are more important than has widely been recognized.[20] One important consequence of the covalent nature of the hydrogen bond is that the energies are not additive. Ab initio calculations have shown that small clusters of hydrogen bonded molecules exhibit both positive and negative cooperativity, depending on the topology of the cluster. In the case of linear clusters of HCN, cooperativity can increase the energy per hydrogen bond by 90% over the energy of a hydrogen bond between only two molecules of HCN.[21] On the other hand, when a donor interacts with three hydrogen bond acceptors, the total energy of the three bonds will be considerably less than the sum of three independent pairwise bonds. This means that adding up hydrogen bond energies, either to estimate binding affinity,[15] or to estimate the energy as a function of internal coordinates of a protein[22] can lead to significant errors when the bonding groups are near one another. As the distance between pairs of hydrogen bonds increases, their energies will be more additive. Some possible consequences of hydrogen bond cooperativity in the binding of acetylcholine to its receptor will be considered below.

Bond Strain

Covalent bonds can be stretched, strained, and twisted at a cost of potential energy, and these are often treated as harmonic oscillators with energies expressed as $\frac{1}{2}\alpha x^2$, where α is a force constant and x is the displacement from the equilibrium position of an extensible coordinate, which could be a distance or an angle. The extent to which the energy of distorting a bond will influence other bonds depends on how rigid the protein is in adjacent regions. A deformation of one bond could spread a considerable distance if this deformation involves displacing a stiff segment. The extent to which bond deformation energies are additive is directly related to how the distortions of the two bonds are correlated. When two force constants in a harmonic potential are altered, it can be shown that the free energy changes will be additive if the two relevant internal coordinates are uncorrelated in position prior to the alterations.[23] In other words, if the stretching or twisting of one bond is independent of the stretching or twisting of another, then the energy change resulting from strengthening or weakening both will be the sum of the energy changes resulting from altering either. More than 90% of the positional disorder in protein crystals can be attributed to lattice motions correlated over distances of about 6 Å or less.[24] Thus, it is likely that groups close together will be correlated so that altering force constants of these contacts will not make additive contributions to the free energy. The small amount of longer range disorder in proteins appears not to be described by correlated rigid body motions within a harmonic potential,[24] suggesting that nonadditivity of changes in bond strain should rarely if ever extend more than 6 Å. Thus, nonadditivity of bond strains is likely to be a factor primarily among neighboring groups, but not among groups separated by greater distances.

For all four of the interactions considered above, nonadditivity is far more likely for pairs of contacts that are near one another (i.e., separated by distances comparable to bond lengths of the contacts themselves). For contacts separated by greater distances additivity is more plausible. Since the multiple contacts that hold a neurotransmitter in a protein binding site are necessarily near one another, it is probably unsound to use Equations 5a and 5b to estimate binding energies and Equation 6 to estimate ΔG_λ. On the other hand, binding sites on different subunits are likely to be separated by much greater distances. With nonadditivity occurring primarily between proximal contacts, the kind of additivity necessary for allosteric theory is still quite plausible. These considerations would tend to support the assumption of additivity in the macromolecular sense. Even if internal protein contacts near a binding site are altered due to short range cooperative interactions with the ligand-protein contacts, these interactions probably will not spread the large distances that separate binding sites on different subunits. Short range deviations from additivity should therefore not make binding energies nonadditive. Equation 3 would still hold, with the provision that ΔG_λ is not defined as in Equation 6, but rather subsumes changes in internal contacts of the protein close enough to the binding sites to be influenced by short range cooperativity with ligand-protein contacts. To illustrate these points in the context of allosteric theory, we will next consider specific scenarios for the AChR.

Nonadditivity at Binding Sites

Given that binding of a ligand to a binding site on a protein implies the formation of new contacts, Equation 9 would suggest that ligand binding cannot accelerate the rate of a conformational transition. This may seem like a surprising statement. But Equation 9 says that the rate is a product of many independent terms, each of which is less than one by an amount specified by the energy, E_i, of the corresponding contact. Within this framework rates can be accelerated by breaking or weakening contacts, but one normally thinks of protein-ligand association as a contact-forming process. Experimentally, it is clear that binding can accelerate the rate of a conformational transition. When agonists bind to the two binding sites of the AChR, the rate of its ion channel opening increases by a factor of 1,000,000.[11] If the model upon which Equation 9 is based is correct, i.e., that a transition is rate limited by the coincident disruption of independent contacts, then somehow a ligand has to weaken intramolecular contacts within the protein that are important to the stability of the closed state of the channel. Contacts that secure ligands to binding sites would have to have significant anticooperative interactions with internal protein contacts that stabilize the closed state of the channel. This is an interesting possibility, leading to some provocative suggestions for ligand-protein interactions.

As discussed above, bond strain interactions can be nonadditive such that straining one bond will strain others. Ligand binding might stress the protein near the binding site so that some of the internal contacts that stabilize the closed state of the channel are stretched closer to their breaking point. In this view, tension would be created in the ligated-closed state when a set of contacts cannot simultaneously assume distances that minimize energy. By displacing an internal protein contact away from its position of minimum energy, the energy required to break the contact will go down, and the rate of channel opening will increase. This would require a simple coupling of ligand-protein contacts with internal protein contacts through the spring-like forces that maintain bond lengths and angles.

One very straightforward way for ligand binding to weaken intrinsic contacts within a protein is to appropriate a hydrogen bonding group that is normally connected to another part of the protein. For example, a number of tyrosine residues in the α subunit of the AChR have been identified in the ligand binding site by affinity labeling and mutagenesis experiments,[7,25] and it is possible that during binding a hydroxyl group from one of these tyrosines forms a hydrogen bond with acetylcholine. If in the absence of ACh the hydroxyl group forms a hydrogen bond with another residue within the protein, then that internal hydrogen bond will have to break when ACh binds (Fig. 4.2). If this internal hydrogen bond contributes to the stability of the closed state of the channel, then the rate of channel opening will be accelerated. If each broken hydrogen bond has an energy of 2 kcal/mol, then Equation 9 gives an acceleration of the rate of opening by a factor of 840 (from breaking one bond at each of the two binding sites). If the energy is 3 kcal/mol, then opening would be accelerated by a factor of 22,000. Thus, this mechanism provides a plausible way in which ligand binding can accelerate channel opening. It should be noted that if forming a hydrogen bond with acetylcholine occurs at the expense of an internal hydrogen bond, then the corresponding contribution to the binding energy will reflect the difference in the two energies. Thus, the 35-fold reduction in binding affinity resulting from the removal of the hydroxyl from tyrosine 190,[26] and 2.1 kcal/mol reduction in binding free energy, may reflect the difference between the energy of the hydrogen bond formed with ACh and the energy of a hydrogen bond broken within the protein.

Some interesting possibilities for nonadditivity can be developed from consideration of hydrogen bond cooperativity and anticooperativity. A hypothetical example is shown in Figure 4.3A, with a chain of four hydrogen bonds involving a bound water molecule, two tyrosines, a serine and a glutamate. If binding neurotransmitter entails displacement of the bound water molecule, without providing a substitute hydrogen bonding partner, then the hydrogen bond chain will be broken (Fig. 4.3B). According to the cooperativity of hydrogen bonds in such linear clusters,[20] the remaining hydrogen bonds of the chain (bonds #1 and #4) will then be weakened. If channel opening requires cleavage of some of those hydrogen bonds, the rate of opening will be accelerated. This is illustrated by the motion depicted on the right side of the drawing in Figure 4.3C, which is shown to be concomitant with the breaking of bond #4.

If a bound water molecule can act as a hydrogen bond donor, another interesting situation can be envisioned. Figure 4.4 starts with a similar linear arrangement of four hydrogen bonds as shown in

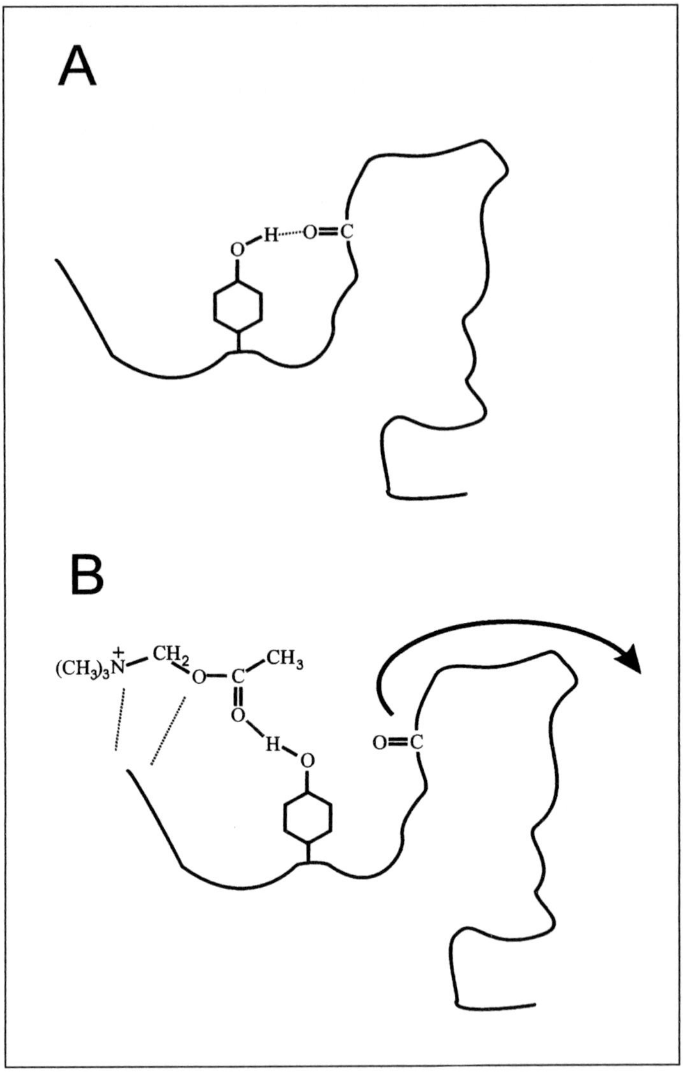

Fig. 4.2. (A) A diagram of a hypothetical structure in the AChR binding site shows a tyrosine hydroxyl forming a hydrogen bond with a peptide backbone carbonyl. (B) Binding of acetylcholine appropriates the hydrogen bond with the tyrosine hydroxyl. Loss of this hydrogen bond within the protein could account for the acceleration of channel opening by agonist binding. An arrow suggests movement of the right loop of the peptide backbone during a conformational change. In this figure and others below, dotted lines signify hydrogen bonds as well as other unspecified noncovalent interactions. Such contacts are presumed to form between various parts of the ACh molecule and other residues of the protein.

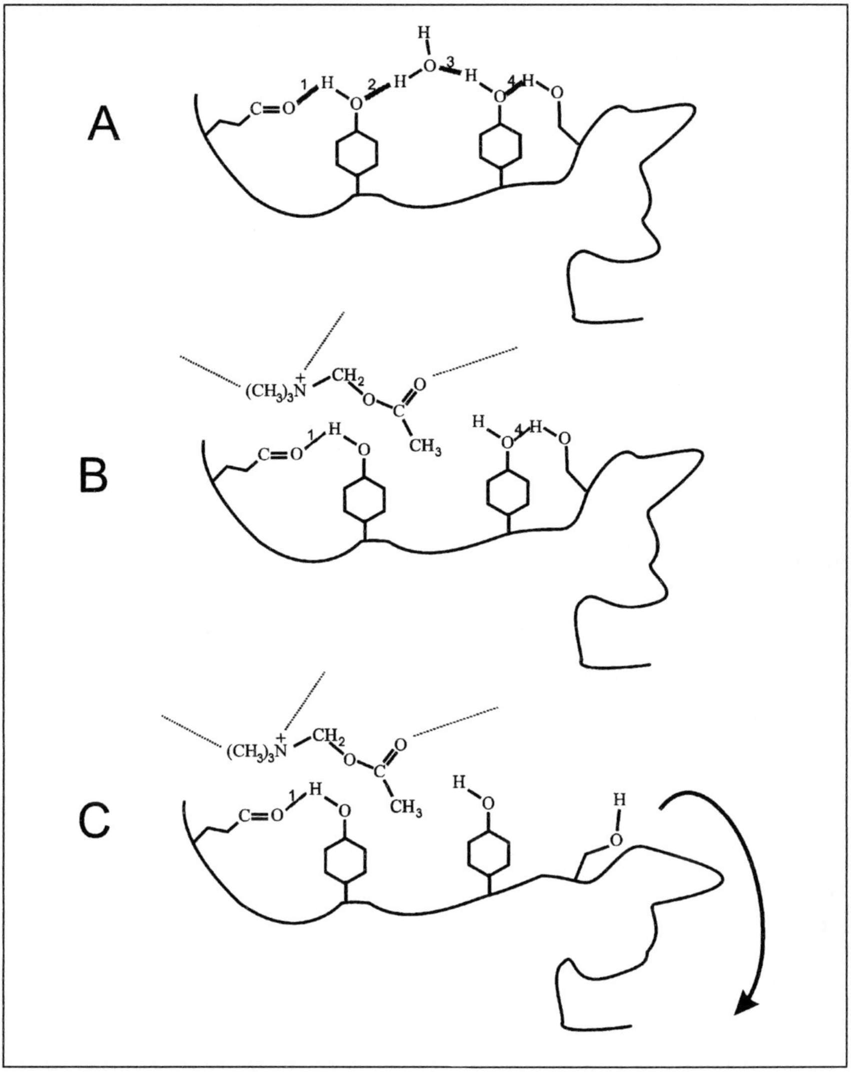

Fig. 4.3. (A) A diagram of a hypothetical structure in the AChR binding site shows a chain of four hydrogen bonds starting with a glutamate carbonyl, to a tyrosine hydroxyl (bond #1), to a bound water (bond #2), to a tyrosine hydroxyl (bond #3), to a serine hydroxyl (bond #4). This chain of hydrogen bonds should form a cooperative unit, such that the bonds are stronger than they would be in isolation (greater strength is indicated by thicker shaded lines). (B) Binding of acetylcholine is shown to displace the bound water without forming hydrogen bonds with the tyrosine hydroxyls. Thus, the second and third hydrogen bonds of the chain are lost. (C) The loss of the cooperative network will weaken the first and fourth hydrogen bonds (thinner shaded lines). Hydrogen bond #4 is shown breaking to permit a conformational transition in the protein as depicted by the displacement of the right loop of the peptide backbone.

Figure 4.3, but here the hydrogen bond chain starts with a donor serine and ends with an acceptor glutamate. In this example, binding of ACh is proposed to involve formation of a hydrogen bond with the bound water molecule, rather than displacement of the bound water shown in Figure 4.3. Since the water already is a hydrogen bond donor (bond #3), forming the new hydrogen bond will place the oxygen atom of the bound water molecule at the center of a star-like arrangement (Fig. 4.4B), which according to theoretical studies of hydrogen bond clusters will lead to negative cooperativity.[20] The hydrogen bond between the acetylcholine and the water will therefore be weak, and anticooperative interactions will weaken bond #3 of the chain between the bound water and the protein. If this hydrogen bond is important to the stability of the closed state of the channel, weakening it will accelerate the rate of channel opening. Furthermore, after bond #3 breaks and the channel opens, the configuration of the remaining hydrogen bonds will once again be cooperative, and that will make the contact between acetylcholine and the protein (via the bound water molecule) become stronger (Fig. 4.4C). This will give the open state a higher affinity than the closed state, thus fulfilling one of the essential tenets of allosteric theory.

Additivity in the M2 Region

M2 is a 19 amino acid segment within the AChR that was originally identified on the basis of hydropathy analysis as a putative membrane spanning domain (see Table 2.1 in chapter 2 and Table 5.2 in chapter 5 of this volume). The first role of M2 to be clearly demonstrated was ion permeation, and the work that established this role also showed that M2 traverses the membrane and lines the aqueous pathway through which ions flow.[7,27] Subsequent studies showed that residues in M2 also influence a number of other receptor functions including desensitization and sensitivity to agonist. As a result M2 has developed into a clinic for studying allosteric mechanisms. Experiments with site-directed mutagenesis have shown that residues in M2 alter desensitization, not only in the nicotinic receptor,[28] but also in the 5-HT_3 receptor[29] and $GABA_A$ receptor,[30] closely related members of the ligand-gated channel superfamily (see chapter 2 in this volume). Furthermore, chemical modifications of substituted cysteines in M2 shift the agonist sensitivity of the receptor.[31] What makes these results interesting is that M2 is so far from the agonist

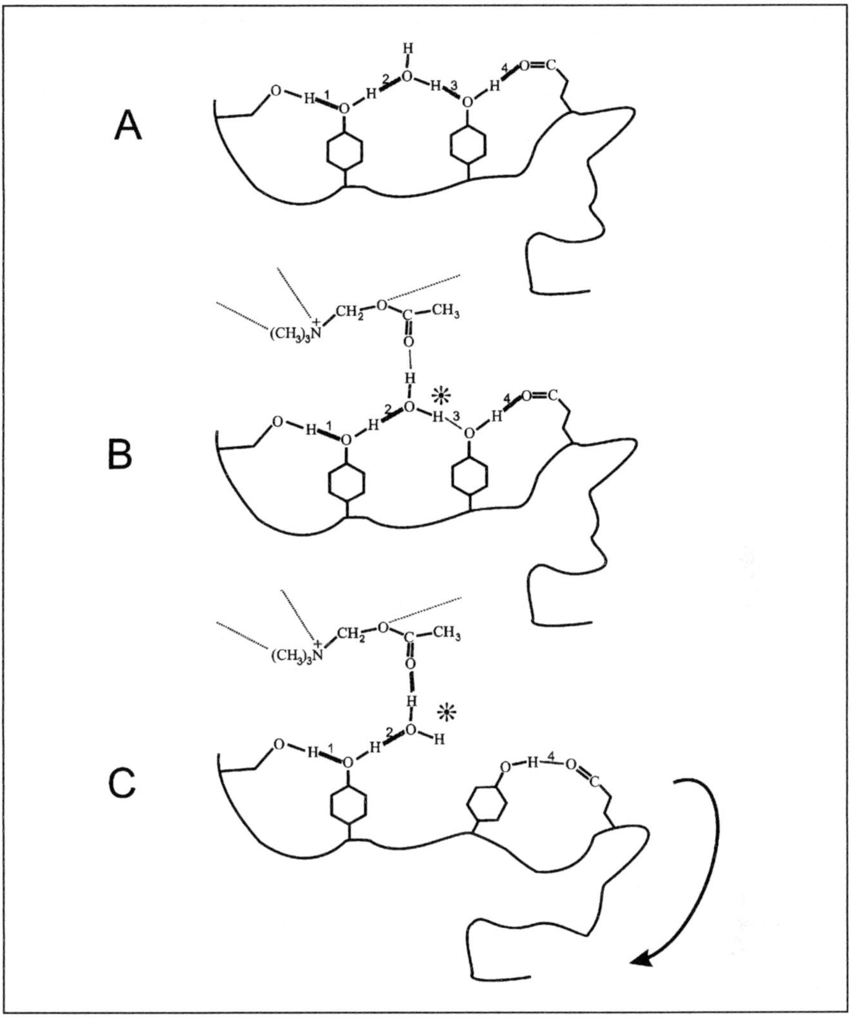

Fig. 4.4. (A) A final hypothetical structure of the AChR binding site shows a similar chain of four hydrogen bonds as in Fig. 4.3, but with the serine and glutamate reversed. As in Fig. 4.3, the thickness of the shaded line indicates the strength of the hydrogen bond. (B) Acetylcholine is now proposed to form a hydrogen bond with the bound water molecule (labeled with ✻). After forming this additional hydrogen bond, the water becomes part of a star-shaped hydrogen bond network that should show negative cooperativity between the two bonds in which the water is a Lewis acid. (C) This negative cooperativity will weaken hydrogen bond #3, allowing it to break more easily so that the protein can undergo the conformational transition. After bond #3 has broken the anticooperative arrangement of hydrogen bonds is lost and the remaining hydrogen bond between the acetylcholine carbonyl and the tyrosine hydroxyl will be part of a linear chain in which hydrogen bonding exhibits positive cooperativity. The bonds will then become stronger and strengthen the attachment of acetylcholine to the binding site.

binding site that it is hard to see how modification of one of these residues can perturb the structure of the binding site. Although a direct effect of an M2 residue on the binding site cannot formally be ruled out, Equation 2 above provides a more plausible explanation for the shift in agonist sensitivity without invoking a structural change in the binding site itself. A shift in sensitivity can then be explained as the result of a change in the equilibrium constant of the allosteric transition.

It is highly significant that residues influencing conformational stability turn up in M2, and this should come as no surprise. As the channel lining segment of the protein, this is a region that must move during gating. It would then make perfect sense to build in important contact-forming residues to hold the channel open and closed, and when one of these residues is modified, the stability of one or more conformations will change.

Recent studies in the M2 region of the AChR produced results very relevant to the questions raised here regarding additivity. Much of the mutagenesis work in M2 has focused on a conserved leucine residue (residue 251 in the mouse AChR α subunit, the 9th residue of M2; see Table 5.2 and Figs. 5.3, 5.5 and 5.6 in chapter 5 of this volume). Replacement of this residue by various amino acids reduces desensitization and increases the sensitivity of the AChR.[28,31-33] Similar results have been obtained for the 5-HT_3 receptor.[29] The heterooligomeric structure of the AChR permits mutations to be made in homologous locations in each of the five subunits, either alone or in combination. When the leucine 251 was replaced either by a serine[33] or threonine,[32] the shifts in AChR sensitivity were nearly perfectly additive. Replacement of one residue shifted the sensitivity by a factor of 10; replacement of additional residues shifted the sensitivity by additional factors of 10.

These results can be explained in terms of contacts formed by the 9th residue of the M2 segment with other unspecified parts of the protein. By interpreting the replacement as the strengthening of a contact in the open state by 1.4 kcal/mole, we can account for the effects of the mutation. If each mutation changes one of the g_{io}s by this amount, T_0 would be increased 10-fold and according to Equation 2 the sensitivity would be reduced proportionally. Furthermore, the strengthening of these contacts would have cumulative effects in slowing the rate of transitions out of the closed state; one result would be a slowing of the rate of desensitization, as observed.

Conclusions

Additivity of energies was considered at two levels. Macromolecular additivity refers to energies that can be measured as protein conformational equilibria and binding affinities. Micromolecular additivity refers to individual contacts between atoms within a protein. Micromolecular additivity implies macromolecular additivity, and macromolecular additivity provides a basis for the theory of allosteric proteins. Thus, the development of ideas regarding additivity at these various levels represents an effort to reduce allosteric theory to a more fundamental level and thus make it more rigorous.

However, the validity of micromolecular additivity receives a mixed review. First of all, the noncovalent interactions commonly thought to govern receptor-mediated signaling processes are often not additive, especially when operating at close quarters. Since the multiple contacts that stabilize the association of a ligand must be near one another, and near some internal contacts of the protein, one should not expect additivity among these contacts. Indeed, the nonadditive properties of hydrogen bonds offer strategies for explaining some of the interesting events that occur during AChR activation. Nevertheless, these violations of additivity arise in such a way as to leave macromolecular additivity intact, and this is because distance is the key. The brief survey of noncovalent interactions conducted here supports a general view that nonadditivity may be quite widespread in proteins but limited to short-range interactions. Since binding sites are separated by large distances, the binding energies may still be additive. Furthermore, the energy changes in internal protein contacts near a binding site can be formally incorporated into the binding energy, leaving contacts distant from the binding sites to generate the intrinsic free energy change of the conformational transition. The finding that interactions distant from the binding sites exhibit pronounced additivity,[32,33] offers hope that other contacts distant from binding sites may also be additive, and independent of the contacts involved in ligand binding. This strengthens the case for in the activation of the AChR as well as other ligand-gated channels.

Acknowledgments

I thank Frank Weinhold for helpful discussions and Shyue-fang Hsu for comments on the manuscript.

References

1. Monod J, Wyman J, Changeux J-P. On the nature of allosteric transitions: A plausible model. J Molec Biol 1965; 12:88-118.
2. Jackson MB. Thermodynamics of Membrane Receptors and Channels. Boca Raton, FL: CRC Press, 1993.
3. Jackson MB. Spontaneous openings of the acetylcholine receptor channel. Proc Natl Acad Sci USA 1984; 81:3901-3904.
4. Jackson MB. Single channel currents in the nicotinic receptor: A direct demonstration of allosteric transitions. Trends Biochem Sci 1994; 19:396-399.
5. Colquhoun D, Sakmann B. Fluctuations in the microsecond time range of the current through single acetylcholine receptor ion channels. Nature 1981; 294:464-466.
6. Changeux J-P. Acetylcholine receptor: An allosteric protein. Science 1984; 225:1335-1345.
7. Karlin A, Akabas MH. Toward a structural basis for the function of the nicotinic acetylcholine receptors and their cousins. Neuron 1995; 15:1231-1244.
8. Unwin N. Acetylcholine receptor channel imaged in the open state. Nature 1995; 373:37-43.
9. Karlin A. Chemical modification of the active site of the acetylcholine receptor. J Gen Physiol 1969; 54:245s-264s.
10. Damle VN, Karlin A. Effects of agonists and antagonists on the reactivity of the binding site disulfide in acetylcholine receptor from *Torpedo californica*. Biochemistry 1980; 17:3924-3932.
11. Jackson MB. Dependence of acetylcholine receptor channel kinetics on agonist concentration in cultured mouse muscle fibres. J Physiol 1988; 397:555-583.
12. Sine SM, Claudio T, Sigworth F. Activation of *Torpedo* acetylcholine receptors expressed in mouse fibroblasts. J Gen Physiol 1990; 96:395-437.
13. Czajkowski C, Kaufmann C, Karlin A. Negatively charged amino acid residues in the nicotinic d subunit that contribute to the binding of acetylcholine. Proc Natl Acad Sci USA 1993; 90:6285-6289.
14. Unwin N. Nicotinic acetylcholine receptor at 9 Å resolution. J Molec Biol 1993; 229:1101-1124.
15. Jackson MB. Perfection of a synaptic receptor: Kinetics and energetics of the acetylcholine receptor. Proc Natl Acad Sci 1989; 86:2199-2203.
16. Jackson MB. On the time scale and time course of protein conformational changes. J Chem Phys 1993; 99:7253-7259.
17. Zwanzig R, Szabo A, Bagchi B. Levintal's paradox. Proc Natl Acad Sci USA 1992; 89:20-22.
18. Barker JA, Henderson D, Smith WR. Three-body forces in liquids. Phys Rev Lett 1968; 21:134-136.

19. Eisenberg D, McLachlan AD. Solvation energy in protein folding and binding. Nature 1986; 319:199-203.
20. Weinhold F. Nature of H-bonding in clusters, liquids, and enzymes: An ab initio, natural bond orbital perspective. Molec Struct 1997; (in press).
21. King BF, Weinhold F. Structure and spectroscopy of $(HCN)^n$ clusters: Cooperative and electronic delocalization effects in C-H···N hydrogen bonding. J Chem Phys 1995; 103:333-347.
22. Brooks BR, Bruccoleri RE, Olafson BD, States DJ, Swaminathan S, Karplus M. CHARMM: A program for macromolecular energy, minimization, and dynamics calculations. J Comput Chem 1983; 4:187-217.
23. Jackson MB. Influence of specific contacts on the stability and structure of proteins. Biophys J 1987; 51:313-321.
24. Clarage JB, Clarage MS, Philips WC, Sweet RM, Caspar DLD. Correlations of atomic movements in lysozyme crystals. Proteins 1992; 12:145-157.
25. Dennis M, Giraudat J, Kotzyba-Hibert F, Goeldner M, Hirth C, Chang J-Y, Lazure C, Chretien M, Changeux J-P. Amino acids of the *Torpedo marmorata* acetylcholine receptor a-subunit labeled by a photoaffinity ligand for the acetylcholine binding site. Biochemistry 1988; 27:2346-2357.
26. Chen J, Zhang Y, Akk G, Sine S, Auerbach A. Activation kinetics of recombinant mouse nicotinic acetylcholine receptors: Mutations of α-subunit tyrosine 190 affect both binding and gating. Biophys J 1995; 69:849-859.
27. Imoto K, Busch C, Sakmann B, Mishina M, Konno T, Nakai J, Bujo H, Mori Y, Fukuda K, Numa S. Rings of negatively charged amino acids determine the acetylcholine receptor channel conductance. Nature 1988; 335:645-648.
28. Revah F, Bertrand D, Galzi J-L, Devillers-Thiéry A, Mulle C, Hussy N, Bertrand S, Ballivet M, Changeux J-L. Mutations in the channel domain alter desensitization of a neuronal nicotinic receptor. Nature 1991; 353:846-849.
29. Yakel JL, Lagrutta A, Adelman JP, North RA. Single amino acid substitution affects desensitization of the 5-hydroxytryptamine type 3 receptor expressed in *Xenopus* oocytes. Proc Natl Acad Sci USA 1993; 90:5030-5033.
30. Zhang H, ffrench-Constant RH, Jackson MB. A unique amino acid of the *Drosophila* GABA receptor influences drug sensitivity by two mechanisms. J Physiol 1994; 479:65-75.
31. Akabas MH, Stauffer DA, Xu M, Karlin A. Acetylcholine receptor channel structure probed in cysteine-substitution mutants. Science 1992; 258:307-310.
32. Filatov GN, White MM. The role of conserved leucines in the M2 domain of acetylcholine receptor in channel gating. Mol Pharmacol 1995; 48:379-384.

33. Labarca C, Nowak MW, Zhang H, Tang L, Deshpande P, Lester HA. Channel gating governed symmetrically by conserved leucine residues in the M2 domain of nicotinic receptors. Nature 1995; 376:514-516.

CHAPTER 5

Molecular Modeling of the Nicotinic Acetylcholine Receptor

Marcelo O. Ortells, Georgina E. Barrantes and Francisco J. Barrantes

Introduction

The Ligand-Gated Ion Channel Common Structure

The ligand gated ion channel (LGIC) superfamily constitutes a major group of receptor proteins.[1] It comprises the nicotinic acetylcholine (AChR, neuronal and muscle-type), 5-HT_3, GABA, and glycine receptors. It is believed, on the basis of sequence similarity, that all members of the family have a similar tertiary structure. Most of the structural information, however, based mainly on data obtained from the nicotinic AChRs, is indirect in that it is inferred from sequence, or biochemical and mutational analyses. The usual picture of the LGIC structure is a pentamer constituted by homologous subunits, each composed of a large extracellular N-terminal domain (which is presumed to bear the f site), four putative helical transmembrane regions (TM), M1 to M4, and a short extracellular C-terminal domain.[2] In the AChRs, those subunits named "α", characterized by the presence of a pair of adjacent cysteines, bear a binding site for acetylcholine (ACh). In the best known of all AChRs, those from *Torpedo* spp. and muscle, there are two of such α subunits and

The Nicotinic Acetylcholine Receptor: Current Views and Future Trends,
edited by Francisco J. Barrantes.

three other different subunits called β, γ (or ε) and δ. As discussed in chapter 3, the probable quaternary (azymuthal) arrangement of the five subunits is α-γ-α-δ-β.

The Binding Site

The two ACh binding sites of *Torpedo* and muscle AChRs exhibit different pharmacological and physiological characteristics, as analyzed in detail by Prince and Sine in chapter 3. It is believed that these differences originate from the topography of each binding site at the interfaces between subunits. The main amino acids known to be involved in the structure of the ACh binding site belong to the α subunit and are, according to *Torpedo* nomenclature, Y93, W149, Y190, C192, C193 (the latter two cysteines are characteristic of the α subunits) and Y198. Replacement of these residues produce, important changes in the binding of nicotinic agonist and dose-response characteristics. It is important to note that all these residues are essentially hydrophobic, as are those involved in the binding of ACh in acetylcholinesterase. They probably contribute to the stabilization of the quaternary ammonium moiety of the agonist. The γ and δ subunits also seem to contribute to shape the binding site or at least the binding pocket of ACh. These residues are γW55 and δW57, both labeled by d-tubocurarine, and δD180 and γD174, that when mutated, greatly alter the affinity for the agonists. It has been proposed that all these residues occur in loops.[3] However, the basis for this proposal is unclear, as many other combinations of secondary structures may be compatible with the labeling and mutation data. Moreover, three rods of electron density visible in electron microscope images of the extracellular region have been interpreted as α-helical structures contributing to the binding pocket, as they are unique to an external cavity in the α-subunits.[4] Higher resolution will be needed to better define the structure of the ligand-recognition region in the presence and absence of ligands.

The Transmembrane Region

In terms of structural interpretations, the best understood region is that of the AChR ion channel proper.[5] This is composed mainly of M2, and biochemical, mutational (see ref. 2 and 5) and electron microscopy[4,6] data have repeatedly confirmed its α-helical nature. The secondary structure of the remaining TMs (M1, M3, and M4) is, however, uncertain. Unwin[4,6] reported the structure of the AChR with the highest resolution so far, at 9 Å, based on cryoelectron microscopy of two-dimensional arrays of a preparation of the protein produced by aging native AChR-rich membranes. A total of five helical structures were resolved in the TM region and these were considered to be the M2 segments of each subunit. A particular feature discovered by this author for the M2s, is that they are kinked in their middle section, approximately at the level of a conserved leucine residue present in all five subunits. This leucine, present in every subunit, consequently forms a ring of residues that putatively is responsible for occluding the central pore in the closed state.

Görne-Tschelnokow et al[7] detected 40% of β-sheet plus turn structure in a preparation obtained by limited proteolysis of membrane-bound AChR and purported to correspond to the AChR TM region, using Fourier transform infrared spectroscopy. Blanton and Cohen[8] have interpreted a series of affinity labeling studies as giving support to the orthodox view that not only M2 but also M1, M3 and M4 are helical. They used 3-trifluoromethyl-3-(m-[^{125}I]-iodophenyl)diazirine ([I^{125}]TID) as a lipophilic photoactivable reagent to label those regions of the AChR that are in contact with the lipid phase of the membrane. M1, M3, and M4 were labeled, and they concluded that for M3 and M4, the labeling pattern is compatible with a helical conformation. They could not, however, assign the pattern obtained for M1 to either a helix or a β-sheet. Moreover, the loop connecting M2 and M3, a region not usually considered to be in the membrane, was also labeled. Thus, an alternative interpretation might be that M2 is not the only helix in the TM and that the α-helix is not the only secondary structure present within the membrane spanning region.

Unwin[4] suggested that the tertiary structure of the TM region might resemble that of the B5 pentamer of the heat-labile enterotoxin.[9] This toxin is a homomeric pentamer with a central channel formed by five α-helices, one from each B subunit. There is also a catalytic A subunit that is not part of the channel. The remainder of the toxin is β-sheet, with the exception of a small helix in the exterior of each subunit. The subunits are joined together by main chain interactions from adjacent β-strands and only neighboring subunits are in contact. Unwin,[4] in the context of LGIC, enumerated as possible advantages for such structure a fixed framework for channels of differing sequence due to nonspecific backbone interactions (versus specific side-chain interactions in the case of α-helices), and stability in the hydrophobic environment.

Molecular Models for the ACHR Structure

The Binding Site

Molecular models for the LGIC binding site are much more difficult to build than those for the transmembrane region (see below), because convincing structural information for this domain is lacking. Residues known to be involved in agonist binding are far apart in the amino acid sequence, and hence it is difficult to predict in which of the three main secondary structure classes they are implicated. This knowledge is essential to begin constructing a detailed three-dimensional model.

Nevertheless, Cockcroft et al[38] built a model of what they believed was a part of the ACh binding site of the AChR. This region was the conserved fifteen residue cys-loop of the extracellular region. Because this cys-loop is present in every known subunit of the LGIC superfamily, and consequently strongly favored by natural selection, Cockcroft et al[38] postulated that this region had to be part of an essential feature of this superfamily. One of these features is the capacity to bind a ligand, and both this capacity and the cys-loop belong to the extracellular domain. Amino acid sequence analysis allowed them to hypothesize which residue within this loop was the key one needed to distinguish among the four known ligands, i.e., ACh, 5-HT_3, GABA and glycine. In spite of these theoretical features, the model had an essential problem: none of the residues of the cys-loop has been recognized experimentally as a part of the binding site. Consequently, the model was never seriously considered. Nevertheless, it has never been proved that this cys-loop is not involved at least in shaping the agonist binding pocket.

Although it has been repeatedly confirmed that there is no significant sequence similarity or evolutionary relationship between the AChR and acetylcholinesterase, Bhat and Taylor[10] proposed a model for the whole nicotinic ACh binding site on the basis of the structure of the latter. This enzyme is clearly related to other esterases and even to thyroglobulin,[11] but not to other ACh-binding molecules. The rationale of Bhat and Taylor[10] was that because both bind ACh, "at least some local sequence and structural similarities between the acetylcholinesterases and the receptors which recognize ACh might be expected." A sequence alignment may have several "interesting" conserved residues; however, without a significant sequence similarity, this is a very weak argument. In the same way, a structural similarity between acetylcholinesterase and the muscarinic acetylcholine receptor might be expected, but this is clearly not the case. Moreover, with the same argument it is possible and a priori equally probable to look for similarities between $GABA_A$ and $GABA_B$ receptors, or between the different types of 5-HT_3 receptors.

In order to construct a model of the agonist binding site on the basis of electron microscope data,[4] Ortells (submitted) applied a combined secondary structure prediction algorithm to the whole LGIC superfamily. He identified four candidate α-helices in the amino acid sequence that may correspond to the three density rods reported to be involved in shaping the binding pocket.[4]

The Transmembrane Region

Hydropathy profiles calculated from sequence data were originally interpreted as indicating a common structural motif of four antiparallel α-helices (M1 to M4) in the AChR (reviewed in ref. 12). This conventional disposition of the TM region of individual subunits could be considered the first accepted model for the region.[12] However, this provides only a very schematic conceptualization and by no means a molecular representation; as such, the chances of being able to validate the model are very low.

Molecular models with atomic level resolution of this region can be separated into two groups: those that consider only the AChR ion channel and those that include what is considered to be the full transmembrane region, from M1 to M4.

"Ion Channel-Only" Models

We do not include in this group molecular models that in fact are excellent pictures made with molecular modeling packages. Though they are much better than schematic representations, they lack a rigorous computational and modeling methodology[12,13] and as such cannot be considered true molecular models. In general terms models of the isolated ion channel domain constructed considering only the M2 segments are also inadequate, since they neglect any participation in the ion channel of the other transmembrane segments, mainly M1 and M3. Even if M1 or M3 do not participate directly in shaping the ion channel proper, such models are incomplete because their electrostatic influence is completely overlooked. Early work on the molecular modeling of the transmembrane region did precisely this, focusing on the ion channel, treating it as a simple cylinder and disregarding the surrounding protein structure.[14-17] All these models were constructed prior to the more recent electron microscope work of Unwin[4,6] that gave support to the existence of only one α-helical structure in the purported ion channel region, attributed to be the M2 segment. On this basis, several recent modeling studies have been restricted to the pore domain.

Sankararamakrishnan and Sansom[18] performed molecular dynamics simulations on bundles of five M2s surrounding a central column of water and with caps of water molecules at either end of the pore. These bundles have been used to explore the effect of water molecules within the pore on helix packing. It was shown that interaction of water molecules with the N-terminal polar side chains lead to a conformational transition from right to left-handed supercoils during these simulations and revealed that the pore formed by the bundle of M2 helices is flexible. These authors extended their modeling on isolated M2 helices to studies on packing interactions.[19] They used simulated annealing via restrained molecular dynamics to build the M2 helices. In the absence of side chain electrostatic interactions, packing of parallel M2s is in accordance with a simple ridges-in-grooves model, resulting in a left-handed coiled coil structure for the bundle as a whole. However, tilting of the M2 helices away from the central pore axis at their C-terminal amino acid residues, or the inclusion of side chain electrostatic interactions, may perturb the ridges-in-grooves packing. Furthermore, in the most extreme cases, right-handed coiled coils are formed. Sansom et al[20] described their results on applying standard molecular dynamics simulations and target restraints derived from Unwin's[4] low-resolution electron microscope images and distance restraints derived from mutagenesis data. They showed that cumulative distortions of backbone (ϕ and ψ) values are responsible for the kinked conformation of M2. The central region of M2 may adopt a more distorted conformation, enabling the ring of conserved leucine residues (one from each M2 helix, see above) to pack together and occlude the pore in the AChR closed state. Extending these studies, Sankararamakrishnan et al[21] compared ion channels formed by isolated M2s from the α7 subunit (which can form homooligomeric AChRs) in the open and closed states. They showed, among other things, that the torsion angles lie within the α-helical region of a Ramachandran plot for both models. However, the distortion is greater in the closed than in the open-state model. This suggests that the shift between closed and open states does not correspond to a mere rigid body movement of the M2 helices, but rather that the region around the conserved ring of leucines may act as a "molecular swivel." With respect to pore dimensions, the minimum radius of the closed-state model was less than 2 Å, and that of the open-state model less than 6 Å. The presence of "bound" water molecules within the pore may reduce these values by up to ~3 Å.

Whole Transmembrane Models

The first models for the whole transmembrane region[22,23] were based on the common assumption at that time that the four putative TM segments of each subunit are α-helical. However, they did not take into consideration the actual dimensions of the channel as measured in electrophysiological experiments.

Ortells and Lunt,[24] testing this traditional model, concluded that an all-helix bundle was an unlikely structure for the TM region. They used models of helical bundles based on four antiparallel helices. This is a well known super secondary structure found, for example, in myohemerythrin. It tightly packs together 4 helices, and was used not only in that work, but also in earlier studies.[22,23] Ortells and Lunt[24] showed that such a structure has severe steric problems: adjacent subunits partially superimpose in space if the dimensions of the pore have to be kept within the range of those calculated from electrophysiological experiments.

Ortells and Lunt[25] have recently proposed that the enterotoxin structure can actually match that of the LGIC. Although the experimentally determined crystal structure of the heat-labile enterotoxin does not correspond to the membrane-associated form of the protein,[9] Ortells and Lunt[25] have recently proposed that the enterotoxin structure can actually match that of the LGIC and be used as a model of the AChR transmembrane region. Their rationale was as follows. First, there is considerable evidence to support the view that the lining of the ion channel in the LGIC is formed by five helices, one from each subunit, and that this is also the case with the toxin channel. One of the surprising features that has emerged from the several hundred proteins whose crystal structures have been solved, is that

Nature has a limited repertoire of tertiary structures.[26,27] Thus, although the toxin and the AChR transmembrane structures are not homologous, it is quite possible that the construction of a pentameric helical channel has been solved in a similar way by nature rather than by the evolution of some entirely different structure. Second, the lack of sequence similarity between LGIC TMs and the enterotoxin B subunits was not an obstacle for modeling purposes. The structure of verotoxin-1 from *E. coli* has been solved[28] and it has a similar tertiary structure to the heat-labile enterotoxin channel. However, there is no sequence similarity, even knowing the matching residues in the three-dimensional structure.[28]

The transmembrane region of the nicotinic AChR was thus modeled in both the closed[26] and open states[29] of the ion channel using the structure of the heat-labile enterotoxin structure as a starting point. As seen below, the closed-state model[26] was obtained using a 5-helix toxin structure with only minor modifications. The open-state model, on the other hand, was obtained by means of major rearrangements of the ion channel region.[29]

The first step in the modeling was the alignment of the TM region of the α7-type AChR and the toxin B subunit sequence, as shown in Table 5.1. Also included in this alignment are one exemplary sequence each of GABA, glycine and 5-HT_3 receptors. The starting point for the alignment was the channel region, since this is the only region of LGIC for which there is a degree of certainty about function, position and secondary structure. The M2s of LGIC were aligned first with helix 2 of the toxin, that is, the helix that forms the channel. As can be seen from Table 5.1, toxin helix 1 and β-strand 1 are not present in the alignment. These form the N-terminal part of each toxin B-subunit, where they cover the external lateral part of the protein. In the alignment there is no counterpart in the LGIC TM of this region. In addition, helix 1 is not present in verotoxin-1. Towards the C-terminal, the toxin has corresponding residues in the LGIC only as far as the middle of M3.

The residues facing the channel lumen were identified by visual inspection of the toxin structure aligning them with those in M2 known to face the lumen on the basis of labeling, mutagenesis, and electrophysiological studies (see reviews in refs. 2 and 5). The alignment of the M2 region provided a framework for the alignment of the remaining TM regions. Further constraints on the positioning of the residues arose from the work of Blanton and Cohen;[8] in

Table 5.1. Sequence alignment of LGIC and heat-labile enterotoxin

	1--------11-- --- ---2 1--------31------- -41 ---- ----5 1- --------61--------
GABRA1	IGYFVIQTYLPCIMTV I LSQ VSFWLNRE SVP ART VFGV TTVLTMT TLS ISARNSLPKVAY A
GLYRA1	MGYYLIQMYIPSLLIV I LSW ISFWINMD AAP ARV GLGI TTVLTMT TQS SGSRASLPKVSY V
SERMA1	PLFYAVSLLLPSIFLM V VDI VGFCLPPD S G ERV SFKI TLLLGYS VFL IIVSDTLPATIG T
ACHCA7	TLYYGLNLLIPCVLIS A LAL LVFLLPAD S G EKI SLGI TVLLSLT VFM LLVAEIMPATSDSV
EHLT	KILSYTESMAG KR EMV I ITF K SGATFQVEVPGSQHI D S QK K AIE R MKDT LR ITY LT ETK IDKLCVW
TOXIN	ßßßßß ßß ßßß ßßßß α α αα α ααα α αα αα αα α αα αα ßßßßßßß
SS	ß2 ß3 ß4 α2 ß5
LGIC SS	ßßßßß ßß ß ßßß α α αα α ααα α αα αα αα α αα αα ßßßßßß
LGIC TM	1111111111 11 111 1 111 11111 2 22 2 222 2 2222 22 222 222 2
LGIC EX	-++-+-+-++ ++ -+- + + -+ -+-++-±- ± - --- + --+ + --++ -+ +-- ++ ---+++±±---±

	71- ------ -81- ---- --- 91 -------101	------ -111--- -- --12 1-----
GABRA1	TAM DWFI AVCY AFVF SAL IEFATVNYFTKR	DRLSRIAF PLLFG IF NLVYWATYLN
GLYRA1	KAI DIWM AVCL LFVF SAL LEYAAVNFVSRQ	DKISRIGF PMAFL IF NMFYWIIYKI
SERMA1	PLI GVYF VVCM ALLV ISL AETIFIVRLVHK	DRLLFRIY LLAVL AY SITLVTLWSI
ACHCA7	PLI AQYF ASTM IIVG LSV VVTVIVLQYHHH	DRLCLMAF SVFTI IC TIGILMSAPN
EHLT	NNK TPNSIA A ISM EN	
TOXIN	ßßßß ß ßßß ßß	
SS	ß 6	
LGIC SS	ßßßß ß ßßß α αα αα ααααααααααα	ααααα α αα α αα αα αα α αα α ααααααα
LGIC TM	33 3 333 3333 333 33 3333333	44444 4 44 4 4444 44 4 444 444444
LGIC EX	± ++ ±±-± ± -+- ++++ +-± ++ -±+++-+	-+++ -+ + -+++ ±+ + -++ +-++--

GABRA1: rat GABA α1 receptor subunit; GLYRA1: rat glycine α1 receptor subunit; SERMA1: mouse 5-HT_3 α1 receptor subunit; ACHCA7: chicken nicotinic α7 AChR subunit; EHLT: *E. coli* heat-labile toxin. TOXIN SS: toxin secondary structure; α: α-helix, β: β-strand, α2, β2-β6: toxin secondary structure nomenclature according to ref. 9. LGIC SS: LGIC secondary structure according to the model. LGIC TM: LGIC orthodox transmembrane regions TM 1 to TM 4. LGIC EX: residue exposure to the lipid environment (TM 1, 3 and 4) or to the ion channel (TM 2) assessed visually from the closed-state model; +, highly exposed; ±, partly exposed; –, hidden.

their studies, four residues from M1 were labeled with a lipophilic probe, and thus are believed to be in contact with or directly accessible from the membrane. The labeled residues were Cys-222, Leu-223, Phe-227, and Leu-228 from the α subunit of *Torpedo* AChR. The loop between M2 and M3 has about 15 residues, with a conserved proline in all LGIC and also another conserved proline in the case of the nicotinic and 5-HT_3 receptors. In the toxin structure, shortly after the channel-forming helix 2 and connected by a three-residue loop, comes β-strand 5. This β-strand does not cross the entire span of the toxin and has to be aligned to the loop between M2 and M3—hence the prediction that part of this loop is within the membrane. In LGIC, the region between the prolines may contain charged residues, which means that if they are within the membrane they are not in direct contact with the lipids. Actually, in the M1 region of the human AChR γ-1 subunit,[31] there is a histidine in the homologous position 28 of Table 5.1. This residue is hidden from the surface in the toxin molecule. Moreover, as we mentioned before, Blanton and Cohen[8] labeled two residues (positions 71 and 72 of Table 5.1) in the loop between M2 and M3, which therefore are probably in contact with the lipids. A visual inspection of the corresponding residues in the toxin protein confirmed that, according to the model, they are perfectly accessible from the membrane (i.e., are visible from the lipid-facing side). These residues belong to the loop connecting β-strand 5 and 6. The latter β-strand, the last portion of the toxin B subunits, is aligned to roughly half of what is considered to be M3. Because it ends in what would be the extracellular side in LGIC, the remaining half of the AChR sequence has to be used to go back to the cytoplasmic side, since the loop connecting M3 to M4 is cytoplasmic.[2]

Owing to its labeling pattern, Blanton and Cohen[8] suggested that M3 is helical; the alignment of its first half with a sequence whose secondary structure is a β-strand is therefore apparently incompatible. However, as can be seen by direct inspection of the toxin and contrary to expectations, consecutive accessible residues are found in β-strands. Nevertheless, because only half of M3 can be homology modeled, and the remaining half has an α-helix compatible labeling pattern, a standard right-handed helix was built for residues 85 to 113 of Table 5.1. This helix was positioned manually with its C-terminal at roughly the same level as the beginning (N-terminal) of M2, and as close as possible to the first part of M3, keeping the labeled residues of positions 87, 90, and 91 visible from the membrane side. Another consideration taken into account when positioning the M3 helix was to "hide" as many of the nonlabeled residues as possible.

There is no toxin counterpart for the M4 region. Hence, this TM region had to be constructed entirely de novo. In early mutagenesis experiments M4 could be replaced by a foreign helix without losing receptor activity.[32] This observation was interpreted as indicating that M4 forms little more than a hydrophobic cover for a part of the remaining structure.[25] Another interpretation is that M4 is probably a structure totally independent of the channel proper. This view may be reinforced by the fact that helix 1 from the heat-labile enterotoxin is absent in verotoxin-1, indicating that it is probably not essential for channel functioning. A further indication that M4 can be considered to be a separate entity is that it is connected to M3 via a rather long and very variable cytoplasmic loop; hence it is not inherently restricted to be close to any other TM segment.

If M4 is a relatively independent domain, its ideal secondary structure would be an α-helix since it does not need backbone interactions with other regions. The helical nature of M4 has a fundamental basis in the labeling pattern obtained by Blanton and Cohen.[8,33] In their studies, M4 was labeled at the homologous positions 107, 110, 113, 114, 117, and 120 of the alignment of Table 5.1. These authors interpreted this pattern as evidence of the helical nature of M4, on the basis that labeled residues have to face the same side in an α-helix structure.

Based on the above considerations, M4 was modeled as a standard right-handed α-helix which was then positioned such that all the labeled residues had to be accessible from the membrane, and that M4 had to cover as many nonlabeled residues on the other TM regions as possible, whilst never burying the labeled ones. Also, M4 was positioned in a way that maximized the interactions with the remaining protein. Finally, the N- and C-terminals of M4 were located approximately at the same level as the N- and C-terminals of M2. The resulting model is shown in Figures 5.1 and 5.2, compared to Unwin's[4] images of the closed-state receptor transmembrane region.

The similarity between the general shape of the model and the images reported by Unwin[4] is very important, because the model was constructed without a priori taking these into consideration. Figure 5.1A shows the electron density maps corresponding to a synaptic view of the TM of *Torpedo* AChR[4] obtained by electron microscopy, compared with the corresponding region of the model in Figure 5.1B. The overall "starfish" shape of the model agrees very well with the electron density profile. In the model, the tips are made of M3 and mainly M4, and the longest distance from the extreme of one tip to the extreme of one of the opposite tips is 94 Å, measured at the backbone level of M4. In the electron micrographs the equivalent distance is ~65 Å, making each tip approximately 15 Å shorter than in the model. The orientation of the M3 and M4 helices at ~52° and 43° with respect to the membrane plane, respectively, might be the reason why they do not become apparent in the electron microscope images. Also, as only M4 is found in the extreme of the tips, the lower electron density may account for the difference in length.

Figure 5.2 shows a lateral view of the AChR TM region. Again, there is overall agreement as regards AChR shape in the images and in the model. In both, the TM has a truncated cone shape, with the shortest diameter in the extracellular leaflet. In the model, this shape

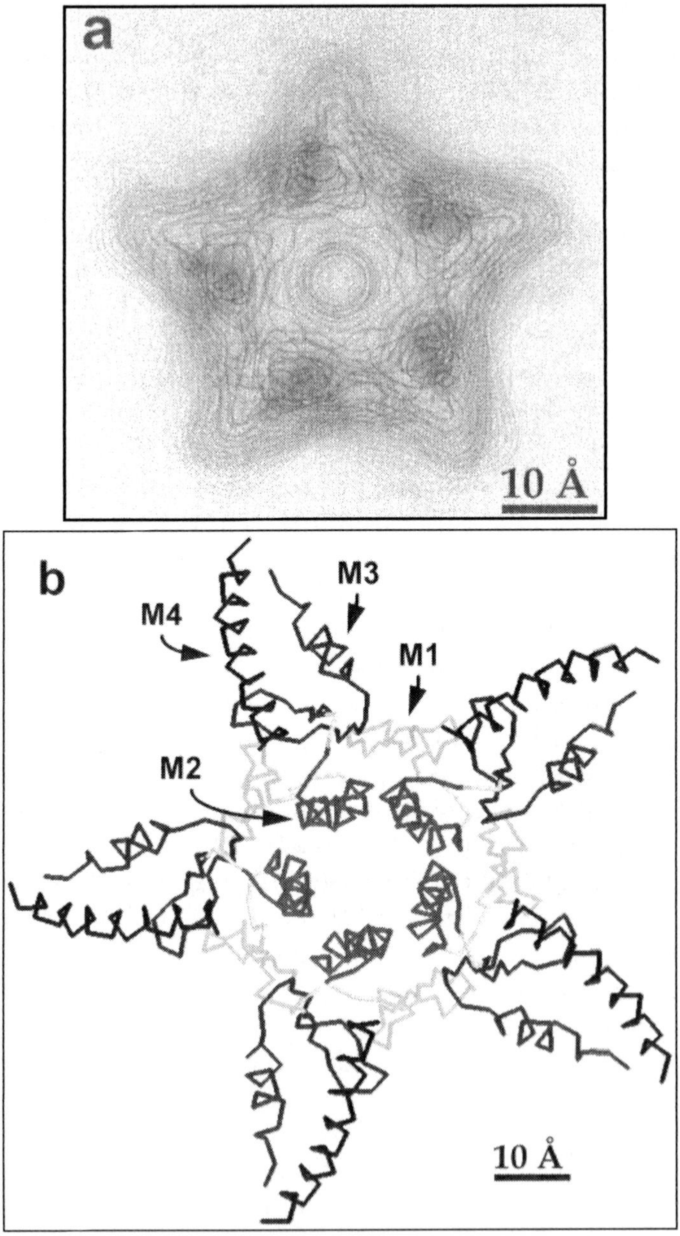

Fig. 5.1. Comparison between Unwin's image[4] (a) and the AChR closed-channel model of ref. 25 (b). Both views are extracellular. Yellow: M1; red: M2, purple: M3; blue: M4; white: connecting loops; green: region previously thought to be in a loop (between M2 and M3) but now part of a β-strand within the membrane in the model. Reprinted with permission from Ortells MO et al, Prot Engng 1996; 9:51-59. See color insert for color representation.

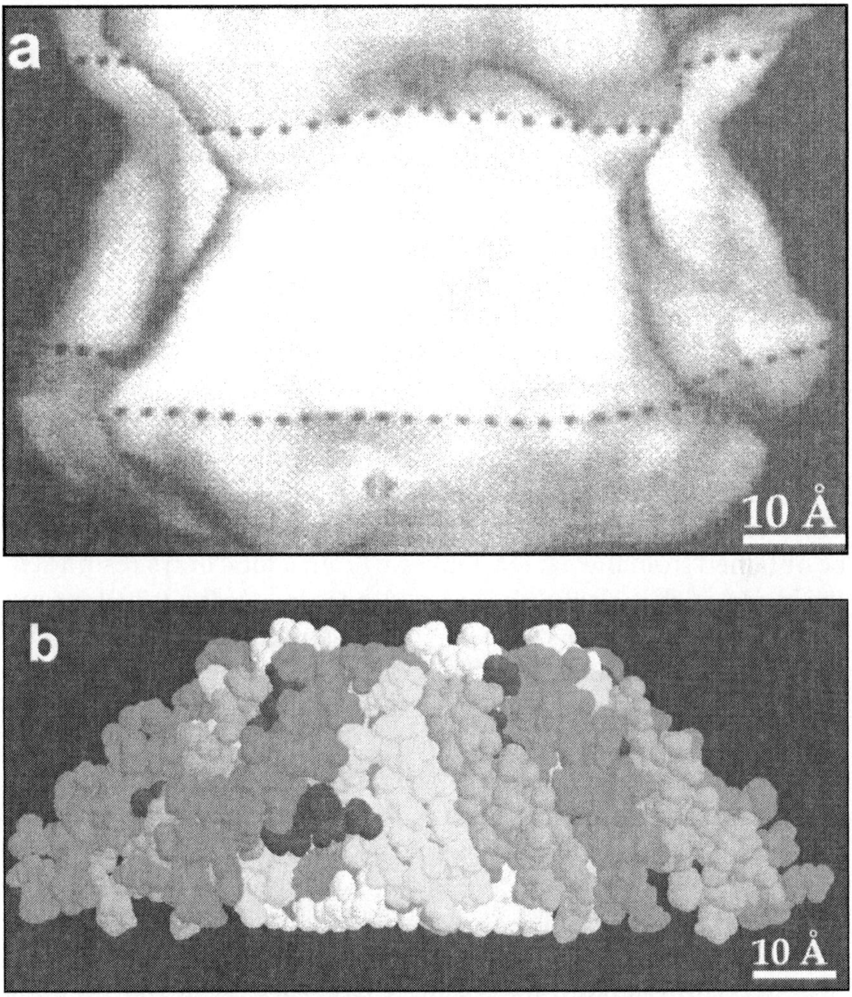

Fig. 5.2. Comparison between Unwin's image[4] (a) and the model (b). Both views are from the side, i.e., as seen from the lipid bilayer. The broken line in (a) represents the boundaries of the bilayer. See Fig. 5.1 for color codes. Reprinted with permission from Ortells MO et al, Prot Engng 1996; 9:51-59. See color insert for color representation.

is mainly due to the inclination of M4. The angle given to M4 in the model was mainly to maximize contact with the rest of the protein. The shape of the AChR TM region is quite asymmetrical, and this may be reflected in the lipid composition in the annulus (see ref. 23 and references therein). The pocket formed between the TM "starfish" tips, more hidden from the bulk lipids, might be the locus for the nonannular sites of cholesterol, as has been previously proposed.[23]

As with the electron microscopy analysis, the TM secondary structure composition data obtained using Fourier transform infrared spectroscopy[7] provide another independent parameter against which the model must be judged. The secondary structure composition measured by this technique in AChR-membranes submitted to controlled proteolysis was 50% α-helix and 40% β-strand plus β turn.[7] The secondary structure composition of the model can be obtained from the data in Table 5.1. From a total of 116 residues in the model of the α7 subunit, 60 (51.7%) are in α-helix, 25 in β-strand (21.55%) and 28 in turns (24.14%) thus, β-strand plus turn makes up 45.7% of the total. Clearly, in support of the model, there is a very good agreement between measured and predicted secondary structure composition.

Because the model of the ion channel domain has basically the same structure as that of the toxin, there was no a priori assumption about whether it represented the open or closed state of the receptor. However, from the final model, it became clear that the structure corresponded to the closed-state (Fig. 5.3). The electron microscope data of the AChR TM in the resting state[4,6] showed that five helices surround the pore and these were assigned to the M2 region. The helices bend near the middle, where they form the narrowest part of the channel. From this point they curve outwards on either side. Unwin[4,1] proposed that the highly conserved leucine (position 49 in Table 5.1) is implicated in narrowing the closed channel in the middle of M2. In the model, the M2 helices are not kinked in the way observed by Unwin[4] but nor are they perfectly straight: they bend slightly. The model concurs with Unwin's images[4,6] regarding the conserved leucines at position 49 of Table 5.1 (Fig. 5.3a,b) and the gradual widening of the channel on either side.[6]

Out of the whole AChR TM region, only the M2 helices have been implicated in the main structural changes occurring during ion channel opening, whereas the external rim is assumed to

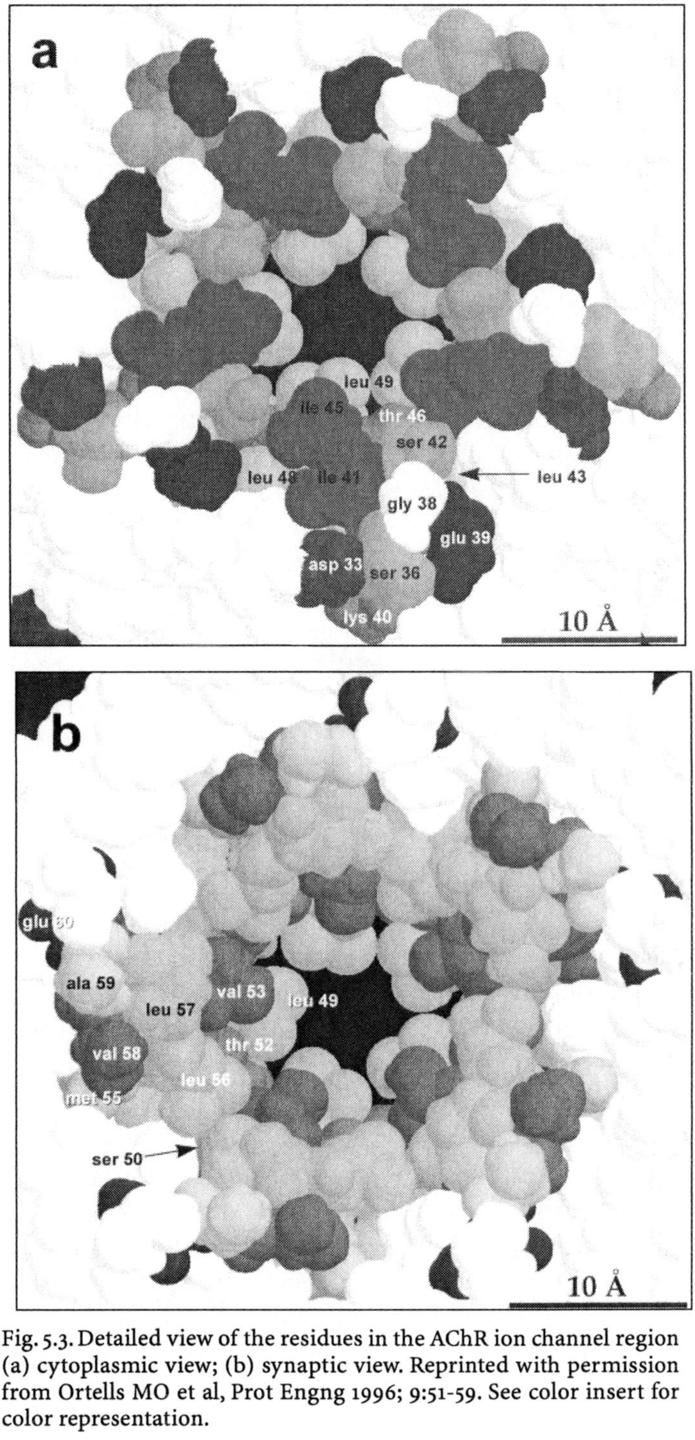

Fig. 5.3. Detailed view of the residues in the AChR ion channel region (a) cytoplasmic view; (b) synaptic view. Reprinted with permission from Ortells MO et al, Prot Engng 1996; 9:51-59. See color insert for color representation.

maintain the same shape in both open and closed states.[6] Ortells et al[29] speculated that the open-state could be modeled by substituting the M2 domains of the closed-state model with appropriately kinked M2-like structures. After searching the Protein Data Bank[34] for suitable helices, all candidate helices were compared in terms of shape with the images reported by Unwin[6] for M2 and those that fitted best were chosen for model building. A copy of each candidate for the new M2 was positioned in the place of the old (closed-state) M2 in such a way that they approximately matched the position of the M2 domains in Unwin's images[6] of the open-state channel. For each of these candidate M2s, an iterative mutational procedure was performed to transform them into AChR α7 M2s. The best candidate was a helix found in aconitase (Brookhaven code 6ACN, residues 110 to 134), and this was used for the construction of the new, open-state M2 domain.

A general schematic view of the open-channel TM region as seen from the synaptic side is shown in Figure 5.4a. One of the most striking differences between the new open-channel model[29] and the closed-state model[25] is that modifications introduced in M2 lead to a slight protrusion of M1 and M3 towards the membrane lipid. The final disposition of the five M2s is asymmetrical. Even though the M2s are identical in sequence and initial conformation, the energy minimization resulted in their backbones occupying similar but not identical positions, and some of the side chains at homologous positions have different accessibilities from the lumen of the channel.

Figure 5.4b shows the estimated molecular surface of the modeled AChR TM region, colored by the electrostatic surface potential calculated by the program Delphi.[35] It can be seen that within the ion channel the electrostatic potential is slightly more negative, as might be anticipated from the known ionic selectivity of the AChR.

Table 5.2 shows a sequence alignment of the four subunits of the *Torpedo* AChR, and the neuronal α7 subunit. The table introduces the numbering that is used in Figure 5.5 and compares it with the conventional numbering used in sequencing and related studies. Figure 5.5 shows an "external" view of the surface of the ion channel with individual side-chain contributions colored differently. An important feature of the model is that residues at position 18 (see Table 5.2) face the channel, albeit with different degrees of exposure due to the asymmetry of the channel model. At that position, the threonine found in the *Torpedo* AChR δ subunit was labeled by the

a

b

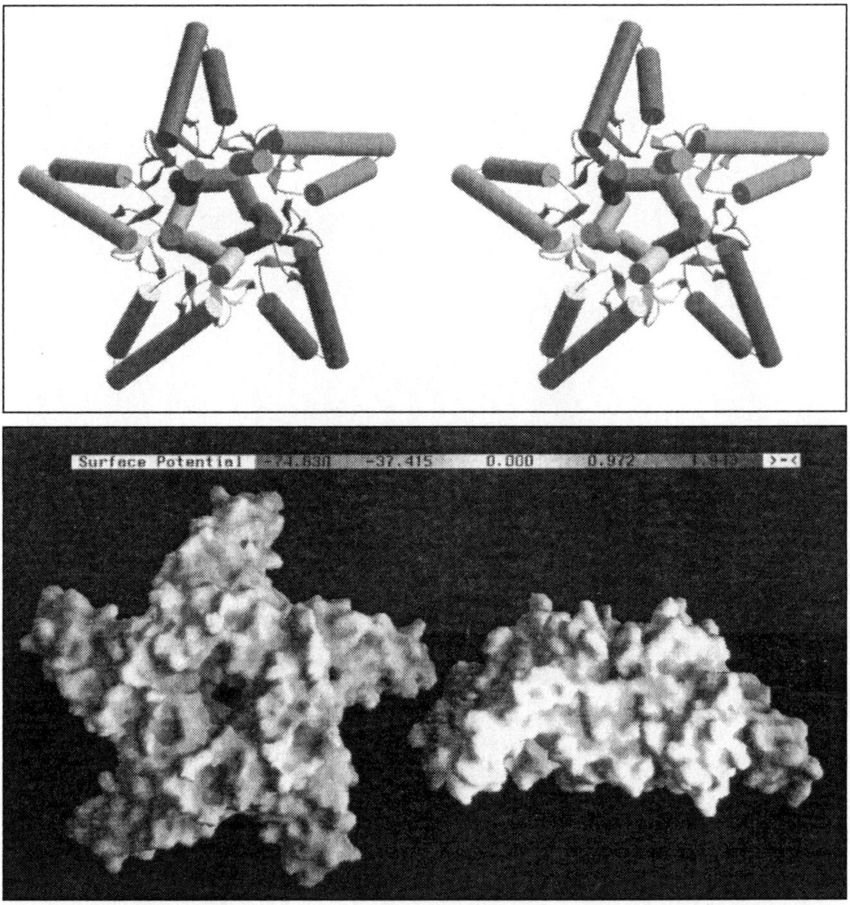

Fig. 5.4. (a) Schematic synaptic view of the whole transmembrane region of the AChR. Each of the five subunits is colored differently. Cylinders are α-helices; flat ribbons are β-strands, and ropes are loops. Generated with the program SETOR (Evans, 1993). (b) Molecular surface generated by the program GRASP, and colored by the electrostatic potential calculated by the program Delphi. Left: synaptic view. Right: lateral (membrane) view. Reprinted with permission from Ortells MO et al.[9] See color insert for color representation.

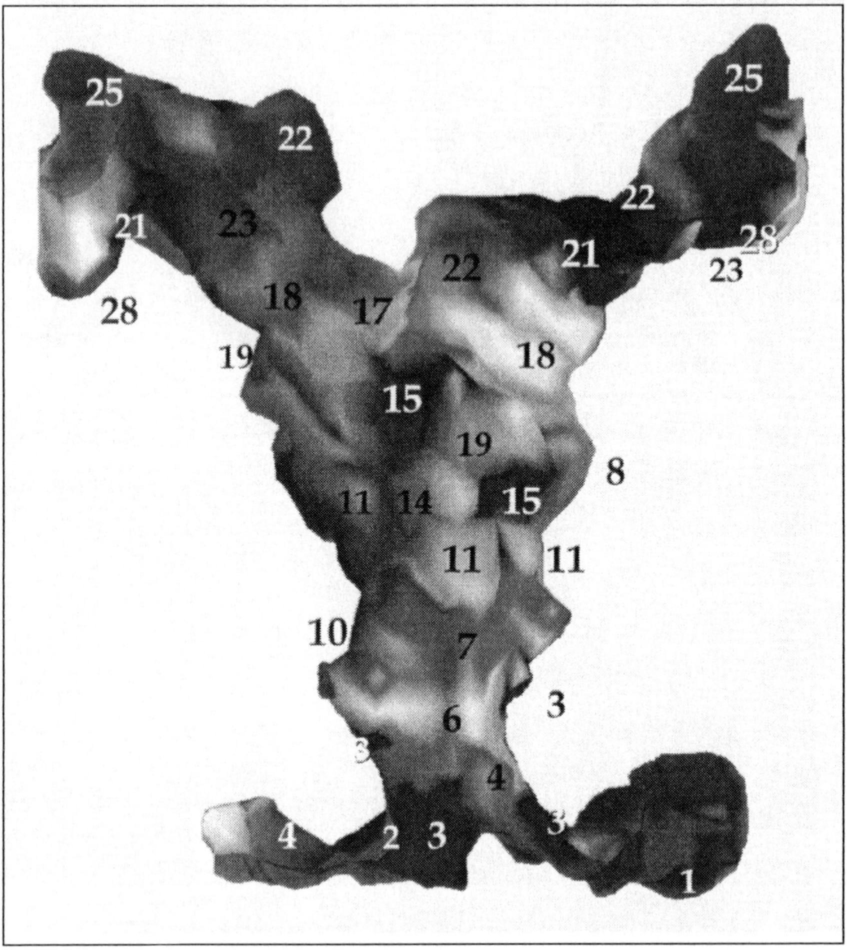

Fig. 5.5. Molecular surface of the ion channel lumen, as viewed from outside the "envelope." Residue numbering 1 to 25 corresponds to that in Table 5.2. Numbers 26, 27 and 28 correspond to the α7 subunit residues Tyr 209 from M1, Met 278 and Ile 279 from M3, respectively. Residue coloring representation is: Red: -5, 2, 8, 14 and +1; Green: -4, 3, 9, 15 and 26; Blue: -3, 4, 10, 16 and 27; Magenta: -2, 5, 21, 17 and 28; Yellow: 1, 7, 13 and 19. Generated by the program GRASP. Reprinted with permission from Ortells MO et al, Prot Engng 1997; (in press). See color insert for color representation.

Table 5.2. Sequence alignment of the M2 transmembrane domain of chick neuronal α7 and Torpedo α1 subunits

n	-5	-4	-3	-2	-1	1	2	3	4	5	6	7	8	9	10	11	12	13	14	15	16	17	18	19	+1
Chicken α7	234	235	236	237	238	239	240	241	242	243	244	245	246	247	248	249	250	251	252	253	254	255	256	257	258
Torpedo α1	238	239	240	241	242	243	244	245	246	247	248	249	250	251	252	253	254	255	256	257	258	259	260	261	262
α7	D	S	G	E	K	I	S	L	G	I	T	V	L	L	S	L	T	V	F	M	L	L	V	A	E
α1	D	S	G	E	K	M	T	L	S	I	S	V	L	L	S	L	T	V	F	L	L	V	I	V	E

Solid black background indicates that residues at that position are exposed to the channel lumen in the open-state model. The light gray background indicates partial exposure and white background nonexposure. The "conventional" sequence numbering of *Torpedo* α1 and chick α7 are also given to facilitate the identification of the residues listed under *n*, an arbitrary numbering adopted for homologous positions along the M2s (residues -2 to +1) and the adjacent N-terminal loop.

photoaffinity label [I^{125}] TID in the presence of agonist.[36] These authors proposed that threonine 23 was labeled from, and was therefore probably pointing toward the lipid interface. They also indicated that this residue "lies on the opposite side of the M2 α-helix from the residues presumed to form the ion channel." As shown, such an explanation is redundant in the model, because the residue is readily accessible from the lumen of the channel.

In the upper region the M2s are quite separate from each other, as they are in Unwin's images,[6] and therefore most of the residues of M2 are "visible" to some degree from the lumen of the channel, with the exception of residues at positions 11, 15 and 19 (Table 5.2). Unwin[6] stated that in the upper half of the channel, the M2 helical segments are sufficiently far apart to permit access to side chains of residues not in M2. In our model, the α7 M1 residue Tyr 209 (residue 26 in Fig. 5.4), and M3 residues Met 278 and Ile 279 (residues 27 and 28 respectively in Fig. 5.4) are accessible from the ion channel, Tyr 209 being at the beginning of M1 and Met 278 and Ile 279 in the second β-strand of M3.

Residues at positions 2, 6, 9 and 10 in Table 5.2 were used as constraints in the construction of the model and are quite exposed. In all cases, however, and because of the asymmetry of the channel, the five homologous side chains are not equally exposed, and are at slightly different levels along the main axis of the channel. It is important to note the relationship between this differential exposure and the different degree of labeling of the four types of subunits (α, β, δ and γ) in the muscle and *Torpedo* receptors, an observation in agreement with the evidence showing that these residues are facing the ion channel in the open-state or in the presence of agonist.

Surprisingly, some residues in the loop between M1 and M2 (positions -5, -4 and -3 in Table 5.2) adopt conformations that constitute a continuation of the orthodox ion channel. Even though they are not closely packed, these residues do shape a well defined path through which ions would pass. Of particular note in this context are the glycines at position -3. It has been shown by Galzi et al[37] that mutations at this region can change AChR ion selectivity from cationic to anionic. These authors speculated that this conversion was not directly related to a change in the channel lumen, but rather to indirect changes in the geometry of the M2s. It is not possible, on the basis of our model, to give a complete explanation of this phenomenon, but the model unquestionably shows that, although not located on the M2, the changes may well directly affect the channel lumen.

Figure 5.6 shows the differences between the M2 structures in the open and closed states, as well as the relative positions of the leucines at position 9 (*Torpedo* αL251) of Table 5.2, i.e., the leucine ring. These residues are believed to close the channel in the closed state.[4] However, Unwin[6] stated that the only differences between the closed and open states in the transmembrane region were at the level of the M2s. The displacement of the open M2s is not the same for the five subunits, as seems to be the case in Unwin's images.[6] All of them appear rotated, but with different degrees of superimposition with the corresponding closed M2s. This difference may be due either to the fact that in the model the superimposition was driven by the external rims, or to undetected asymmetries in Unwin's images. The latter can be explained by the fact that the images were obtained using an averaging method that applies a five-fold symmetry operation.

Acknowledgments

This work was supported by grants from Fundación Antorchas, Argentina, to M.O.O., and from CONICET, Argentina, and the European Union grant No. CI1*-CT94-0127, to F.J.B.

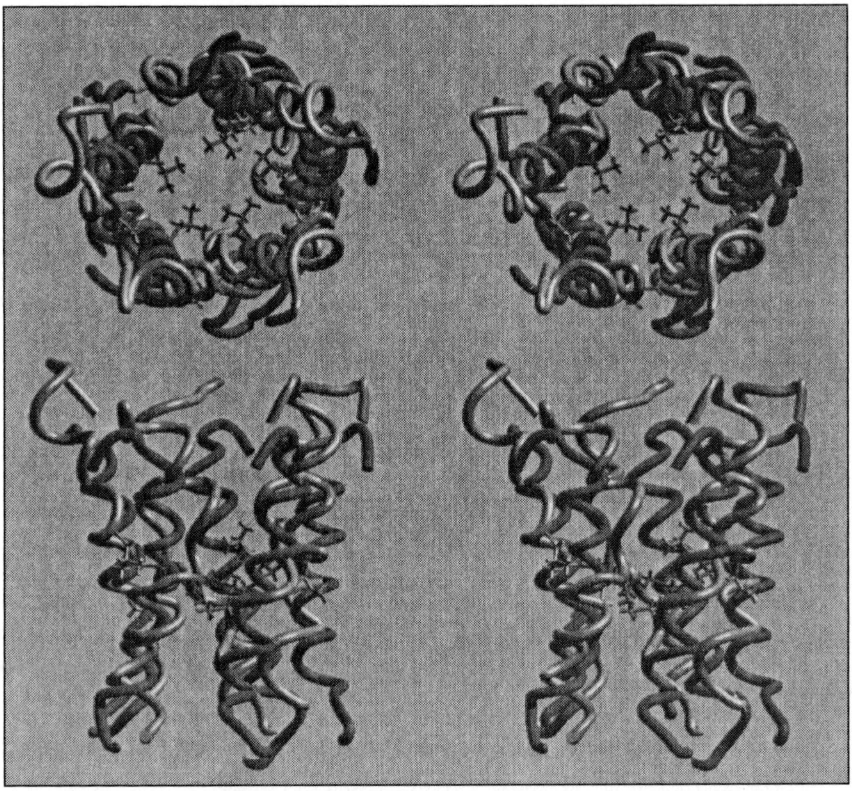

Fig. 5.6. Schematic stereo view of superimposed M2 helices in the closed (green) and open (gray) states (rms value of 6.9). Above: cytoplasmic view; below: lateral view, with the cytoplasmic half of the channel in the upper part of the figure. The side chain of the leucines at position 9 of Table 5.2 are also displayed to show the differences in their position in the open and closed states, respectively. Reprinted with permission from Ortells MO et al.[29] See color insert for color representation.

References

1. Unwin N. Neurotransmitter action: Opening of ligand-gated ion channels. Cell 1993; 72:31-41.
2. Karlin A. Structure of nicotinic acetylcholine receptors. Curr Opin Neur 1993; 3:299-309.
3. Galzi J-L, Revah F, Bouet F et al. Allosteric transitions of the acetylcholine receptor probed at the amino acid level with a photolabile cholinergic ligand. Proc Natl Acad Sci USA 1991; 88:5051-5055.
4. Unwin N. Nicotinic acetylcholine receptor at 9 Å resolution. J Mol Biol 1993; 229:1101-1124.
5. Bertrand D, Galzi J-L, Devillers-Thiéry A et al. Stratification of the channel domain in neurotransmitter receptors. Curr Opin Cell Biol 1993; 5:688-693.

6. Unwin N. Acetylcholine receptor channel imaged in the open state. Nature 1995; 373:37-43.
7. Görne-Tschelnokow U, Strecker A, Kaduk C et el. The transmembrane domains of the nicotinic acetylcholine receptor contain α-helical and β structures. EMBO J 1994; 13:338-341.
8. Blanton MP, Cohen JB. Identifying the lipid-protein interface of the *Torpedo* nicotinic acetylcholine receptor: Secondary structure implications. Biochemistry 1994; 33:2859-2872.
9. Sixma TK, Pronk SE, Kalk KH et al. Crystal structure of a cholera toxin-related heat-labile enterotoxin from *E. coli.* Nature 1991; 351:371-377.
10. Bhat AA, Taylor EW. Structural evaluation of distant homology—a 3-d model of the ligand-binding domain of the nicotinic acetylcholine-receptor based on acetylcholinesterase—consistency with experimental-data. J Mol Model 1996; 2:46-50.
11. Hucho F, Järv J, Weise C. Substrate-binding sites in acetylcholinesterase. Trends in P Sciences 1991; 12:422-426.
12. Stroud RM, McCarthy MP, Shuster M. Nicotinic acetylcholine receptor superfamily of ligand-gated ion channels. Biochemistry 1990; 29:11009-11023.
13. Akabas MH, Kaufmann C, Archdeacon P, Karlin A. Identification of acetylcholine-receptor channel-lining residues in the entire m2 segment of the alpha-subunit. Neuron 1994; 13:919-927.
14. Furois-Corbin S, Pullman A. The effect of point mutations on energy profiles in a model of the nicotinic acetylcholine receptor (AChR) channel. Biophys Chem 1991; 39:153-159.
15. Furois-Corbin S, Pullman A. Theoretical study of potential ion-channels formed by bundles of α-helices. Partial modeling of the acetylcholine receptor channel. In: Pullman A, Jortner J, Pullman B, eds. Transport Through Membranes: Carriers, Channels and Pumps. The Netherlands: Kluwer Academic Publishers, 1988:337-357.
16. Hilgenfeld R, Hucho F. Properties and problems of the helix-M2 model of the acetylcholine receptor-ion channel. In: Pullman A, Jortner J, Pullman B, eds. Transport Through Membranes: Carriers, Channels and Pumps. The Netherlands: Kluwer Academic Publisher, 1988:359-367.
17. Sansom MSP. The roles of serine and threonine side chains in ion channels: A modeling study. Eur Biophys J 1992; 21:281-298.
18. Sankararamakrishnan R, Sansom MSP. Water-mediated conformational transitions in nicotinic receptor M2 helix bundles—a molecular-dynamics study. FEBS Letters 1995a; 377:377-382.
19. Sankararamakrishnan R, Sansom MSP. Modeling packing interactions in parallel helix bundles—pentameric bundles of nicotinic receptor m2 helices. Biochimica et Biophysica Acta 1995b; 1239:122-132.

20. Sansom MSP, Sankararamakrishnan R, Kerr ID. Modeling membrane-proteins using structural restraints. Nature Struct Biol 1995; 2:624-631.
21. Sankararamakrishnan R, Adcock C, Sansom MSP. The pore domain of the nicotinic acetylcholine receptor: Molecular modeling, pore dimensions, and electrostatics. Biophys J 1996; 71:1659-1571.
22. Cockcroft VB, Lunt GG, Osguthorpe DJ. Modeling of binding sites of the nicotinic acetylcholine receptor and their relation to models of the whole receptor: In: Kay J, Lunt GG, Osguthorpe DJ, eds. Protein structure, prediction and design. Biochemical Society Symposium (57), London: Portland Press, 1990:65-79.
23. Ortells MO, Cockcroft VB, Lunt GG, Marsh D, Barrantes FJ. The nicotinic acetylcholine receptor and its lipid microenvironment. In: Pullman A, Jortner J, Pullman B, eds. Membrane proteins: Structures, interactions and models. Dordrecht: Kluwer Academic Publishers, 1992:185-198.
24. Ortells MO, Lunt GG. The transmembrane region of the nicotinic acetylcholine receptor: Is it an all-helix bundle. Receptors and Channels 1994; 2:53-59.
25. Ortells MO, Lunt GG. A mixed helix-beta sheet model of the transmembrane region of the nicotinic acetylcholine receptor. Prot Engng 1996; 9: 51-59.
26. Efimov AV. Favoured structural motifs in globular proteins. FEBS Letts 1993; 334:253-256.
27. Efimov AV. Super secondary structures involving triple-strand β sheets. Structure 1994; 2:999-1002.
28. Stein PE, Boodhoo A, Tyrrell GJ et al. Crystal structure of the cell-binding B oligomer of verotoxin-1 from *E. coli*. Nature 1992; 355:748-750.
29. Ortells MO, Barrantes GE, Wood et al. Molecular modeling of the nicotinic acetylcholine receptor transmembrane region in the open state. Prot Engng 1997; 10:511-517.
30. Shibahara S, Kubo T, Perski HJ, Takahashi H, Noda M, Numa S. Cloning and sequence analysis of human genomic DNA encoding γ subunit precursors of muscle nicotinic receptor. E J Biochem 1985; 146:15-22.
31. Tobimatsu T, Fujita Y, Fukuda K. et al. Effects of substitution of putative transmembrane segments on nicotinic acetylcholine receptor function. FEBS Lett 1987; 222:56-62.
32. Blanton MP, Cohen JB. Mapping the lipid-exposed regions in the *Torpedo californica* nicotinic acetylcholine receptor. Biochemistry 1992; 31:3738-3750.
33. Bernstein FC, Koetzle TF, Williams GJB et al. The protein data bank: A computer-based archival file for macromolecular structures. J Mol Biol 1977; 112:535-542.

34. Nicholls A, Sharp K, Honig B. Protein folding and association: Insights from the interfacial and thermodynamics properties of hydrocarbons. Proteins: Struc Func Genet 1991; 11:281-296.
35. White BH, Cohen JB. Agonist-induced changes in the structure of the acetylcholine receptor M2 regions revealed by photoincorporation of an uncharged nicotinic noncompetitive antagonist. J Biol Chem 1992; 267:15770-15783.
36. Galzi J-L, Devillers-Thiéry A, Hussy N et al. Mutations in the channel domain of a neuronal nicotinic receptor convert ion selectivity from cationic to anionic. Nature 1992; 359:500-505.
37. Cockcroft VB, Osguthorpe DJ, Barnard EA et al. Modeling of agonist binding to the ligand-gated ion channel superfamily of receptors. Proteins: Struc Func Genet 1990; 8:386-397.

CHAPTER 6

Ion Conduction Through the Acetylcholine Receptor Channel

Alfredo Villarroel

Introduction

Ion conduction describes the process by which ions cross a channel pore and interact with the channel wall. During its journey the individual ion diffuses along the bulk solution to the entrances of the channel where part of its hydration layer is lost. As it enters the pore the interaction with the pore wall compensates for the lost hydration energy. One can imagine the cation spending more time near comfortable groups such as oxygen in hydroxyl and carboxyl groups, where the electrostatic interactions are favorable. So large is the variety of ions that permeate the acetylcholine (ACh) receptor (AChR), that this protein constitutes the perfect system to study ion conduction. In addition, many features of the AChR have been well known for some time from the biochemical, pharmacological, and structural point of view, as analyzed in other chapters of this volume. From biochemistry we know the amino acid composition and the most likely secondary structure of the channel region. From pharmacology we know how blockers interact with the pore. From structural studies we know the channel dimensions and pore size. Knowledge attained from these areas is being readily integrated into electrophysiological measurements.

The techniques used to study ion conduction have, through experience and evolution, become more sophisticated and progressively more expensive. Postsynaptic potentials were used in the 1950s. With

The Nicotinic Acetylcholine Receptor: Current Views and Future Trends,
edited by Francisco J. Barrantes.

the development of the voltage-clamp technique it was possible to study the synaptic currents associated with the nicotinic responses. Since Neher and Sakmann developed the patch-clamp technique,[1] it has been possible to observe the "dancing" of a single molecule in what is called single-channel recording. This period could be called *Ars antiqua* (old art) of ion conduction. What marked the *Ars nova* (new art) period was the cloning of the receptor and the subsequent understanding of its structure and function. The AChR was the first ion channel to be cloned.[2,3] The cloning of the receptor per se was not the turning point, but it opened the possibility of changing its structure by point mutations, and testing the influence of such mutations on ion conduction. In order to carry out these studies, it was necessary to develop expression systems suitable for electrophysiological measurements. Miledi had already introduced the *Xenopus* oocyte preparation to express ion channels.[4] He injected mRNA from denervated cat muscle into oocytes to correctly determine the conductance of the fetal type AChR, 29 pS. The technique was soon improved.[5] The oocyte technique became particularly efficient when using cRNA instead of total mRNA. A recently developed technique uses optical methods to study ion conduction.[6] The determination of fractional-calcium current (P_f), enables us to clearly "see" the influx of Ca^{2+} when a mixture of ions is flowing into the cell.

The protein that forms the AChR of both muscle and *Torpedo* electric organ is composed of five subunits, four of them differing, in the stoichiometry $\alpha_2\beta\gamma\delta$. Furthermore, the muscle receptor has an extra subunit, ε, which replaces the γ subunit during muscle development after birth.[7,8] Each subunit is organized into four amphipathic segments which are thought to cross the membrane four times. The second transmembrane segment (M2) forms the wall of the pore (see chapter 5 for details). Since this chapter considers ion conduction, most of the amino acid residues discussed will be located in the M2 segment.

The molecular composition of the neuronal AChR is different from that of muscle and *Torpedo* electric organ.[9,10] I will mainly discuss, but not exclusively, the receptors from muscle and *Torpedo*. The neuronal AChR will be treated extensively in chapter 7 of this volume. Today the distinction between muscle and *Torpedo* AChR on the one hand, and neuronal receptor on the other hand has become very clear. Many of the controversial findings have a simple explanation if we consider that these two receptor classes differ in their permeation pathway.

This chapter is arranged according to ion types, and what can be learned from studying their permeation properties. Organic ion permeation, which is described in section two, reveals the gross architecture of the AChR pore, in particular the mean diameter of the narrow region. Alkali cation permeation exposes the intimacies of the selectivity filter, and provides a molecular insight into selectivity mechanisms. Section four concerns anions, which are not permeant. From these we have learned about the charge discrimination mechanism in the AChR as well as in several other members of the superfamily of ligand-gated channels (see chapter 2 in this volume). Section five concerns divalent ions, Ca^{2+} in particular. This section is essentially descriptive, since divalent ion conduction mechanisms are still not completely elucidated. The last section delineates future directions in the field of AChR ion conduction.

Permeation of Organic Ions

Organic ions are certainly not physiological, except perhaps ammonium. Nevertheless, physiologists have used them as probes of the permeation pathway of ion channels. Organic ions have a clear advantage for this purpose; they can be chemically designed in a variety of sizes and textures. The favored organic ions have been ammonium derivatives. In particular, quaternary-ammonium ions that have a permanent charge and an hydrophobic surface (such as tetra-ethyl-ammonium, TEA) have been utilized. The size of an organic ion is defined in terms of the van der Waals radii of individual atoms, and its molecular structure.[11] Since the bond length and angles are uniform among atoms in organic compounds, the calculation of molecular dimensions is straightforward.[12] Another advantage is the absence of hydration. The study of organic ion conduction through the AChR revealed important features of the permeation pathway.

The AChR is permeable to a variety of organic ions.[13,14] In early studies performed on the endplate of frog muscle, Hille and colleagues determined the permeability of 40 different organic ions.[14] Permeability for organic ions was dependent on ion size. Ions which cannot dfit in a sectional area of 6.5 x 6.5 Å were virtually impermeant. From this finding we learned that the narrow part of the pore must have a cross section of at least that size. Among permeant organic ions, permeability was higher for smaller ions. Single-channel conductance was also higher for smaller ions.[15] The mechanism of organic-ion permeation, according to what has been observed, follows the excluded-volume principle. The idea is the following: When an organic ion moves through the pore, the center of the ion is not accessible to the complete pore because due to its size the ion can not get close to the wall. Therefore the free space for ion movement is the difference between the pore size and the ion size. In the extreme case of an ion of almost the same size as that of the pore, there will be no free space for ion movement, and the friction with the wall will make the ion impermeant. According to this view the free space for ion movement has a cross-sectional are of $\pi \cdot \left(r_{pore} - r_{ion}\right)^2$. The relationship between permeability (P) and ion size (rion) is:

$$P = k' \cdot \left(r_{pore} - r_{ion}\right)^2 = k \cdot \left(1 - \frac{r_{pore}}{r_{ion}}\right)^2 \qquad (1)$$

were k is a proportionality constant. The excluded-volume can also be applied to the control ion, Na^+ for example.[16] Then the permeability ratio can be written as:

$$\frac{P_X}{P_{Na}} = \left(\frac{r_{pore} - r_X}{r_{pore} - r_{Na}}\right)^2 \qquad (2)$$

Where r_X is the radius of the organic ion used. This relationship predicts an $r_{Na} = 2.1$ Å, close to the value of 2.15 Å for the hydrated radius of Na^+ in water.[16]

Factors Determining Permeability

In addition to pore size, there are other factors that determine the permeability of organic ions, which are either not totally clear, or have not been systematically studied. Unsaturated compounds, perhaps due to the planarity of the double bond, are more permeable than the saturated compounds of equivalent size.[14] It is possible then that the cross section of the pore has an oval shape that facilitates the passage of planar molecules.

Another factor that promotes permeability is the formation of hydrogen bonds. This could be an explanation of why ligand-gated channels are highly permeant to ammonium, in comparison to alkali metal cations of equivalent size. The ammonium permeability

P_{NH_4}/P_{Na} is 1.79 and 2.1 in frog muscle[14] and rat recombinant receptors,[17] respectively. Hydroxyl groups, such as those from serine and threonine residues in the wall of the pore, have the remarkable property of acting as both donor and acceptor.[18,19] A hydroxyl group from the wall of the AChR may use one of their two lone electron pairs to bind ammonium. When a hydrogen bond is formed, the distance between the two atoms involved is shorter than the sum of the individual van der Waals distances. In this way ammonium can get closer to the pore wall. In agreement with this explanation, the single-channel conductance for ammonium was larger than the value that its size would predict.[20] This behavior is called "supra-IA" in Eisenman's terminology and indicates a particular orientation of the ligands in the selectivity filter (see next section).

Tris Crossing the Pore

With the use of recombinant DNA techniques it was possible to modify the pore wall in order to study in greater detail the factors determining ionic permeability. One of the organic ions that has been studied in greatest detail is $Tris^+$. Tris is an organic molecule with dimensions of 6.0 x 6.9 x 7.7 Å (Fig. 6.1A). It has a hydrocarbon surface, and an ammonium group able to make hydrogen bonds. Lester and his group made point mutations of several residues in the M2 segment of each subunit of the mouse AChR, and determined the permeability to Tris of the mutated channels.[21] With this mutation procedure they made channel pores with different textures. The mutant dS2'F had considerably lower permeability for Tris, as did several other mutants at the 2' position (see chapter 5 in this volume). It did not matter in which subunit the mutations were made; only those that changed the hydroxyl ring (position 2') affected Tris permeability (Fig. 6.1). Furthermore, the permeability to Tris was correlated with the hydrophobicity of the side chain. This correlation was more apparent when only mutations at the hydroxyl ring were considered, consistent with the ion-wall interaction at the narrow region of the pore. No clear correlation was found between the permeability and volume of the amino acid side chain.

Molecular modeling of the conduction portion of the AChR (see chapter 5 and references therein) indicates that Tris is one of the largest molecules that can cross the pore. In an ion channel whose pore wall is made of α-helical structures, there is a close correlation between the pore size and the number of subunits.[22] Since the ion

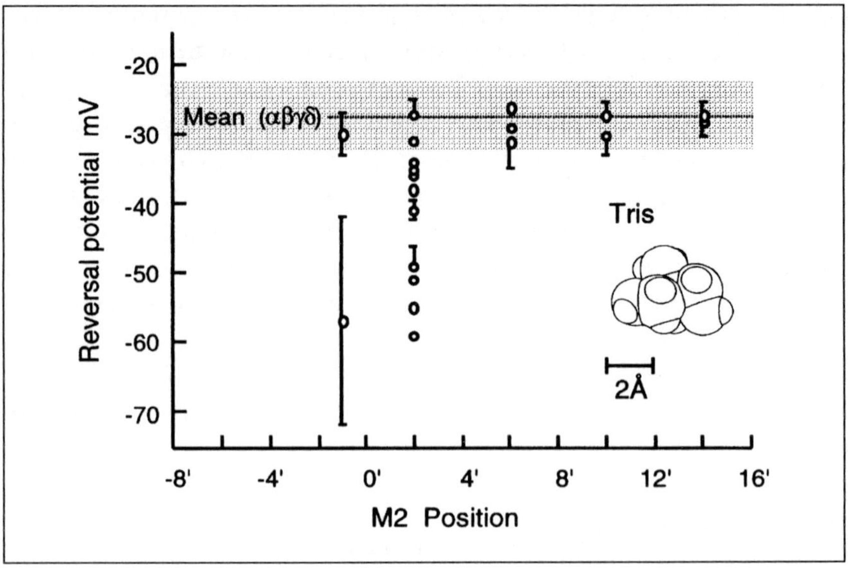

Fig. 6.1. Tris permeability of AChR mutated in the M2 segment. Modified from Cohen BN et al, J Gen Physiol 1992; 99(4):545-72. © 1992, The Rockefeller University Press.

transport occurs between subunits, a hexamer such as a gap junction has a larger pore size than a tetramer, such as a sodium channel. The package of five α-helices in the AChR leaves a space of ~44 Å2, sufficient to fit a Tris molecule. Thus, the pentameric structure of the AChR assures the pore size.

The Kyoto group, working on Torpedo AChR, contemplated the problem of organic ion permeation from a different perspective.[23] They investigated the changes in pore size caused by mutations at the intermediate ring. To do this, they determined the permeability of mutated channels to organic ions of several sizes. One interesting observation that they made was that the permeability for organic ions can be improved. The AChR mutant γQ250N had higher permeability for organic ions than wild-type channels. In this particular mutant channel the pore size increased from 7.43 Å to 7.98 Å. The permeability to Tris can be increased even more upon the replacement of γQ250 by alanine, which has an even smaller side chain. This suggests that both the hydroxyl ring and the intermediate ring restrict Tris movement during permeation.

Length of the Narrow Region

The length of the narrow region can be estimated from streaming-potential measurements. When identical solutions bath each side of the membrane, and an impermeant nonelectrolyte is added in one side, an osmotic gradient is established. As a consequence, water flows through the channel. This water drags ions along the channel, producing a potential that can be measured, named streaming potential. There exists a relationship between the streaming potential and the number of water molecules (N) inside the channel:[24,25]

$$V_{streaming} = N_w \frac{RT}{F} \varphi_s \frac{n_s}{n_w} \quad (3)$$

Where R, T, and F have their usual meaning, φ_s corresponds to the molal-osmotic coefficient for the nonelectrolyte, and n_s and n_w are the number of moles of nonelectrolyte and water respectively. For the AChR expressed in BC3H-1 cells the streaming potential was 2.4 mV/osmolal, corresponding to 6 water molecules.[26] This indicates that the narrow region of the pore is 3-6 Å long, i.e., the length of one α-helix turn. In other words, at most one ring of side-chain residues form the narrow region of the pore.

The Vestibules

Organic ions that are impermeable are also useful probes of the channel vestibules. These ions act as pore blockers and reveal cavities along the conduction pathway. QX-222, a derivative of lidocaine, blocks the AChR channel in a voltage-dependent fashion. As it blocks the pore from the external side, QX-222 experiences approximately 60% of the applied voltage. This means that the energy of the ion at the blocking site increases by 0.6·kT each time that the voltage increases by 25 mV. A methylated nitrogen in one extremity of the QX-222 molecule forms a polar end, and a dimethylated aniline group forms a hydrophobic end.

QX-222 interacts with two adjacent turns of α-helices of the AChR. The charged ammonium head of QX-222 interacts with the side chain residues at position 6', which corresponds to the next α-helix turn after the hydroxyl ring. There the dissociation rate of blockade decreases as the site is made more polar by replacing hydrophobic with hydroxylated residues. In contrast, at the next position toward the extracellular side, position 10', the dissociation rate is increased by adding hydroxylated residues. Thus the charged head of QX-222 interacts with position 6' and the hydrophobic tail with position 10' of the M2 segment.[27] Position 6' is at the end of the external vestibule. Since the blockade by QX-222 has a voltage dependence of 78%,[28] only 22% of the voltage would drop across the rest of the pore. In particular, a small fraction of voltage drop is assigned to the narrow region.

Usually an organic molecule has not a single, but several conformations, each of them giving the ion a different shape and size. Among the many conformations that an organic ion can adopt, some of them will fit better in the vestibules. Skok applied these principles to the neuronal nicotinic AChR.[29,30] He postulated that the organic ions with the largest number of conformations that fit in a defined cavity will have the largest blocking potency. This hypothesis implicitly assumes the absence of specific interactions, such as hydrogen bonds, between the blocker and the pore wall. Thus for a given cavity size the potency of blockade, experimentally measured as the EC_{50}, will be proportional to the number of conformations that fit in it. Since some conformations, due to their lower energy, occur more frequently, this proportion was also taken into account. In order to obtain the best correlation between EC_{50} and fraction of conformations in the minimal size, the size of the cavity should be 7.0 x 8.4 x 9.0 Å at the entrances of the external vestibule, and a cross-sectional area of 5.8 x 8.0 Å at the narrow region.[30] As we have seen in the previous chapter, computer-aided molecular modeling can be used to learn about the shape and dimensions of the AChR channel. In a molecular model of the α7 neuronal AChR, Sankararamakrishnan et al[31] estimated a minimum pore size of 6 Å for the open AChR channel and less than 2 Å for the closed channel. For the Torpedo AChR channel, the model predicts a minimum diameter of ~6.5 Å. The location of the narrow region thus differs between the two channels. For the *Torpedo* channel the minimum is located at the hydroxyl ring, consistent with selectivity studies.[17,20] In the case of neuronal α7, however, the minimum is located more toward the extracellular side.

The Pore

The pore architecture that emerges from the study of organic ions is not that different from those determined from electron microscopy.[32] The external vestibule is quite large and is deep enough to accommodate the QX-222 molecule. The narrow region of the pore is very short, and is probably formed by a single ring of side-chain residues, one from each of the five subunits that form the AChR. The diameter of the narrow region estimated from organic-ion permeation is about 7.4 Å. Structural data suggest a pore size of 10 Å.[29] The difference is probably due to a layer of water molecules at the surface of the protein.[33] The first layer of water is known to be tightly bound to the protein surface. The exchange rate of this first layer of water is slower[34] and it may therefore be considered structural water.

Alkali Cations

The AChR has been known to be permeant to monovalent cations since the beginning of the century. Overton in 1902 noticed that isotonic Li^+ can replace sodium and restore normal synaptic transmission in the neuromuscular junction.[35] Later, Fatt found that Na^+, although necessary to produce propagated responses, is not required for the ACh-induced depolarization of the endplate.[36] An analysis of the endplate potentials showed that ACh produces a short-circuit action making the endplate permeant to all ions.[37] One of the first quantitative determinations of monovalent ion selectivity in endplate AChR channels was done by Takeuchi and Takeuchi.[38] They measured end-plate currents in the sartorius muscle using the voltage-clamp technique. They found that the reversal potential of the endplate currents varies according to the logarithm of the external Na^+ concentration, as is expected from a Na^+ electrode. This was also true for external K^+. These observations demonstrated that the endplate channels were permeable to both Na^+ and K^+. Afterwards, and for no particular reason, the problem of ion conduction at the endplate channels was left dormant for about twenty years.

Permeability Selectivity

Selectivity refers to the idea of how the channel "prefers" certain ions and rejects others. It always implies a comparison between ions. One indicator of selectivity is the permeability ratio determined by the reversal potential, which is usually measured under bionic conditions. Thus the permeability ratio defined as:

$$\frac{P_X}{P_Y} = \frac{[Y]}{[X]} \exp\left(-\frac{FV_{rev}}{RT}\right) \tag{4}$$

is a measure of selectivity. Another determination of selectivity is the conductance sequence or, equivalently, the conductance ratios. Usually these two measurements do not provide the same result. Eisenman and Horn[39] showed that in a very simple conduction model where the channel conductance follows a Michaelis-Menten relationship, the permeability ratio is related to the maximal conductance as:

$$\frac{P_X}{P_Y} = \frac{K_Y}{K_X}\frac{G_X^{\max}}{G_Y^{\max}} \tag{5}$$

where K_x and K_y are the half saturating concentration. As Equation 5 indicates, even in the simple Michaelis-Menten approximation, permeability ratios are different from conductance ratios, except in the particular case where both ions bind to the channel with the same affinity.

Hille and his group determined the permeability ratios for several monovalent cations in 1980.[40] This was important because the mechanism of conduction could be inferred from the studies of a series of alkali cations. Usually physiologists avoid nonphysiological ions, considering it irrelevant to learn from something that is not normally present in the biological preparation. Hille and his group, without such prejudices, used the vaseline-gap technique to investigate the ion selectivity of endplate AChR channels. Using this technique, it is possible not only to uniformly control the membrane potential, but also to perfuse the internal compartment of the muscle. In this way they were able to have an accurate control of both internal and external solutions.

From reversal potential measurements, the selectivity sequence for Cs>Rb>K>Li was found for the AChR channel, which is exactly the sequence for ion mobility in water. The endplate AChR channel from frog-muscle was pictured as an aqueous pore where ions move freely in a solution-like medium. These measurements were consistent only with those made in nonmammalian muscle.[40] As Table 6.1 shows, mammalian muscle AChRs have different selectivity sequences.

In addition to alkali cations, early studies found that Tl^+ was also very permeant ($P_{Tl}/P_{Na} = 2.51$). The Tl^+ permeability was even higher than that predicted from the water mobility ($P_{Tl}/P_{Na} = 1.49$). This would indicate that polarization effects also play a role in ion

Table 6.1. Selectivity by permeability ratios

Preparation	Li	K	Rb	Cs	Reference
Rat myotubes		0.6			Ritchie and Fambrough, 1975
Eel electroplaque		1.1			Lassignal and Martin, 1977
Chick embryo		1.47	1.52	1.91	Huang et al, 1978
Mouse diaphragm	1.1	0.8			Linder and Quastel, 1978
Frog muscle	0.9	1.1	1.3	1.4	Adams et al, 1980
Toad muscle	0.62	1.11		1.60	Quartararo et al, 1987
Torpedo recombinant	0.75	1.05	1.10	1.14	Konno et al, 1991
Mouse, recombinant	0.98	1.16	1.31	1.22	Cohen et al, 1992
Rat recombinant		1.20	1.24	1.23	Villarroel and Sakmann, 1992

Values normalized with respect to Na.

transport.[41] Reversal potentials measured in mixtures of Na^+-Tl^+ and Na^+-K^+ varied monotonically with the mole fraction of the mixture. This observation indicates that no significant double occupancy occurs in the AChR. If ions are likely to interact inside the pore, a minimum in the reversal potential would occur at a particular mole fraction, as has been shown for voltage-dependent Ca^{2+} channels.

Permeability measurements in recombinant mammalian and *Torpedo* AChRs confirmed previous observations. In *Torpedo* receptors the permeability, measured from reversal potentials, follows mobility in water.[42] Cs^+ was the most permeable ion, and at the same time the one with greater mobility. Rat recombinant receptors were slightly more permeable to Rb^+ than Cs^+, consistent with the idea that ion flow is restricted for large alkali cations.[17,20] Smaller alkali cations probably move through the receptor pore together with their complete hydration shell.[17,40]

Conductance Selectivity

Conductance selectivity requires the determination of single-channel conductance in different ions. The recombinant Torpedo AChR has a single-channel conductance of 87 pS in 100 mM KCl and in the absence of divalent cations.[43] The adult recombinant receptor from rat has a comparable conductance, 104 pS, when measured under similar conditions.[44] The fetal type receptor from rat has a lower conductance, 68 pS. In the presence of divalent cations the fetal and adult rat receptors has a conductance of 43 and 63 pS respectively.[44] Native rat receptors have comparable conductances.[45]

The single-channel conductance sequence for alkali metal cations of the native toad muscle AChR was found to be K>Cs>Na>Li.[46] The recombinant *Torpedo* AChR follows the sequence K>Rb>Cs>Na>Li.[42] The fetal type AChR from rat muscle has an equivalent conductance sequence.[20] The conductance for ammonium is comparable to that of Cs^+ in spite of the small size of the former. This is probably due to the ability of ammonium to form hydrogen bonds that allows ammonium to get closer to the ligand group.[20]

Selectivity Filter

The selectivity filter is the region of the pore that selects ions. This region may not necessarily be the narrowest part of the pore. However, in the particular case where ions are selected according to size, the selectivity filter corresponds to the narrowest region. Since the M2 transmembrane segment probably forms the wall of the pore,[43,47] the amino acid residues that form the selectivity filter are expected to be located there.

We know about the selectivity filter in the AChR from point mutation studies. In the M2 segment there are abundant hydroxylated residues (threonines and serines) which confer its amphipathic character. Hydroxylated residues play a key role in ion conduction. The replacement of a serine by alanine in the M2 segment of both α- and δ-subunits causes a decrease in the single-channel conductance for outward currents.[47] This was one of the first observations that hydroxylated residues are probably facing the pore, and are therefore likely to participate in the ion selection process. However, the narrow region of the pore was still unknown. Part of the reason was that several mutations performed by Lester's group at Caltech failed to yield currents, and it was concluded that "data are not available to decide this point."[27]

Mutations in the threonine αT264 of the rat AChR (αT241 in *Torpedo* channel) lead to the identification of the selectivity filter. The replacement of αT264 and equivalent residues in the other subunits caused a decrease in the single channel conductance of the AChR.[48,17] This indicated that the αT264 and equivalents all face the pore, as the presence of a bulkier residue perturbs ion transport. Mutations that decrease the side chain volume of the residue at that position increased the channel conductance. By some mechanism the ion flow depends on the space available in the pore. A bulky group decreases this space, and leads to a decrease in conductance. The fact that it is possible to increase the conductance is very important because it indicates that there is no other region in the pore that restricts the ion flow. In other words, there is no other region which acts as a rate-limiting step in the transport. These studies indicate that the hydroxylated ring determines ion flow, but they do not prove that this is the selectivity filter. In order to examine selectivity, a sequence of ions must be examined. The single-channel conductance of a channel mutated in the hydroxyl ring for ions in the series of alkali metals was affected in different ways for each ion. Conductance for small ions was barely affected, as if small ions were insensitive to variations in the pore size. Conductance for large ions, on the other hand, was strongly affected. In a channel containing a large side-chain residue, the conductance for Cs^+ was strongly reduced.

Selectivity in the AChR therefore is composed of two mechanisms: (1) a free movement of small ions according to their water mobility and (2) a restriction in flow for large ions. The pore size of the AChR based on organic ion permeability is 7.4 Å diameter. The pore size based on electron microscopy studies is 10 Å.[29] The difference is probably due to immobile water around the pore. Water is known to exist in an immobilized state on the surface of proteins.[49] The restricted permeability for large ions could be due to difficulties in carrying water that adheres to the wall of the pore.

How Much Does a Filter Filter?

The selectivity filter restricts the ion flow, in particular for large ions. We may then ask ourselves how much of the total pore resistance is due to the filter. The total channel resistance is composed of two resistors in a series, that of the vestibules and that of the filter. I assume that no filter is a glycine filter where the side chain of all hydroxylated residues is replaced by a proton. The conductance of such a mutated channel (αT264G*βG278*γT275G*δS279G) for Cs^+ is 78 pS,[50] i.e., larger than that of the wild-type channel (68 pS). The contribution to the filter is 12% of the total pore resistance (Fig. 6.2).

The Anionic Rings

The discovery of the anionic ring was one of the greatest steps towards understanding ion conduction and cationic selectivity of the AChR, and later extrapolated to other ligand-gated channels. Keiji Imoto[43] noticed that as Ca^{2+} was lowered below 1 mM, *Torpedo* and bovine AChRs showed differences in conductance, in addition to gating differences. The conductance of the *Torpedo* channel was larger than that of the bovine channel. Using chimeras, he was able to localize the region responsible for this difference in the M2 segment. A chimera carrying the M2 segment of *Torpedo* receptor will produce channels with larger conductance, no matter what the origin of the rest of the subunit is.[43]

Once the region responsible for the conductance was localized at the M2 segment, Imoto searched for the amino acids involved. At the two ends and in the middle of the M2 segment, he recognized clusters of negatively charged amino acids, which later become designated as the "anionic rings." Point mutation performed in the charged amino acids and single-channel measurements of the mutated channels showed a proportionality between conductance and the number of negative charges. The magnitude of inward currents (at -100 mV) was proportional to charges in the extracellular ring, but not that of the outward currents. Conversely, the magnitude of outward currents (at +100 mV) was proportional to the number of charges in the intracellular ring. The magnitude of both inward and outward currents was proportional to the number of charges in the intermediate ring. Hybrids of *Torpedo*-mouse AChR corroborated these findings.[51] Species differences in single-channel conductance were explained by the charge distribution flanking the M2 transmembrane domain. Two recent independent studies using chimeras of the γ- and ε-subunits of mouse[52] and rat[43] also confirmed these findings.

In addition, the anionic rings were responsible for the sidedness of Mg^{2+} blockade. In the wild type receptor intracellular Mg^{2+} mainly blocks outward currents, and extracellular Mg^{2+} mainly inward currents. Neutralization of the charges in the intracellular ring abolish the intracellular Mg^{2+} blockade, but not that of the extracellular side. Neutralization of the charges in the extracellular ring does just the opposite. As neutralization of the charges in the intermediate ring abolishes both intracellular and extracellular Mg^{2+} blockade, the charge dependence on conductance and the sidedness of Mg^{2+} block-

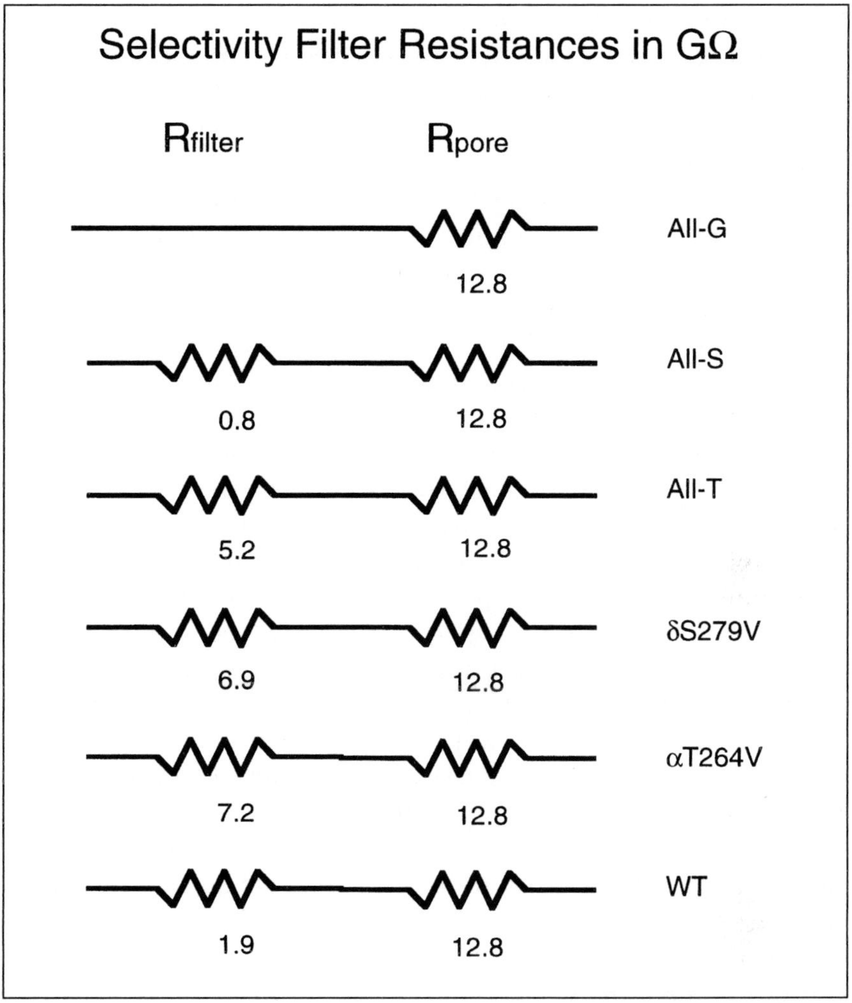

Fig. 6.2. Contribution of the selectivity filter to the total resistance of the pore. The conductance of the mutant AChR in which the hydroxyl ring was replaced by glycine residues, was considered as the pore contribution to the conductance.

ade are exactly what one would expect if the rings act by a fixed charge effect concentrating cations at the entrances of the pore. The electrostatic effect of the charges can be neutralized either by removing the amino acid with point mutations, or by increasing the ionic strength, either of which diminishes electrostatic interactions.

In a second mechanism, also consistent with existent observations, the charged residues provide binding sites for cations actively participating in ion translocation.

Models of the Rings

Dani, without knowing the existence of the anionic rings, modeled the effect of the charges in the vestibule.[53] The shape of the vestibule was considered in the potential, calculated from Gauß's law. In addition to the potential due to the shape of the vestibule there is also a diffuse-double layer potential produced by both the vestibule and the ion atmosphere surrounding them. In this way Dani proposed a Poisson-Boltzman equation that contains the shape of the vestibules. Compared with a planar geometry, a funnel-shaped vestibule produced a potential that decreased faster over distance. Dani allowed ions to bind to the charges of the vestibule, and in doing so, reduced the cross-sectional area. In other words, ions were allowed to partially block the channel. The reduction in the cross-sectional area was included in the entry rates in a two-barrier one-site Eyring model. The vestibule model did not consider the exact position of the charges. The findings of Imoto, assigning positions to the three anionic rings of the AChR, were published two years later.[54] Nevertheless, important as they may be (as shown below), the exact position of the charges does not seem to be crucial.

When B. Sakmann and I tried to predict the effect of point mutations in the fetal form of the rat AChR, we noticed that it was not possible to describe the conductance-concentration using a simple Michaelis-Menten saturation.[17] The conductance at low concentration was always larger than the predicted value. A simple approach is to assume that the local concentration is larger than the bulk concentration, due to the presence of the charged rings. Then we modified the Michalis-Menten formalism to include local concentration instead of bulk concentration. The local concentration was estimated as:

$$[Cs]_{local} = [Cs]_{bulk} \cdot \exp\left(-\frac{F\phi_{surface}}{RT}\right) \tag{6}$$

where $[Cs]_{local}$ and $[Cs]_{bulk}$ are the local and bulk concentration respectively. $\phi_{surface}$ is the dimensionless surface potential due to the charges represented by the Grahame equation,[55] also referred to as the Gouy-Chapman approximation:

$$\sigma^2 = 2\varepsilon\varepsilon_0 kT \sum_j C_j \left[\exp\left(-z_j \phi_{surface}\right) - 1\right] \tag{7}$$

Using this simple model, it was possible to satisfactorily describe the conductance-concentration curves.[17]

The role of the ring appears to be to concentrate cations near the channel entrances. Yellen and his group, however, found evidence against this purely electrostatic mechanism.[56] Neutralization of the charge in the internal ring (αE262Q) produced a decrease in channel conductance that was not completely attenuated at high ionic strength. Mutations in the extracellular ring produced even more unexpected results. Introduction of positive charges in the extracellular ring (αD238K) decreased the channel conductance independently of the ionic strength.[56] If the role of the rings were to concentrate cations near the entrances of the channel by an electrostatic mechanism, it is expected that a high ionic strength would screen this effect. Yellen proposed that cations bind to the charges, decreasing the effective concentration of binding sites. Mutations that alter the charges can also decrease the effective concentration of binding sites accordingly. With this additional assumption the conductance-concentration curves for mutant and wild type were satisfactorily described (Fig. 6.3). Yellen's model did not consider the energy profile of the pore, as Dani's model did. The beauty of Yellen's model resides in the correct prediction of the relationship between conductance and charge in the vestibule for a series of mutant AChR channels.

The rings also act as binding sites. In the *Torpedo* AChR the intracellular, intermediate and extracellular rings have -3, -4, and -3 electronic charges, respectively. Since the single-channel conductance is maximal at pH 7, and not different from that at pH 9, the acidic residues seem to all be deprotonated.[57] Assuming that all of these residues can form binding sites, the AChR channel can contain up to nine ions in the pore, and still remain neutral, i.e., no ion repulsion would occur. This corresponds to an average ionic concentration of 1 M inside the channel. Interestingly, these ions, except perhaps for

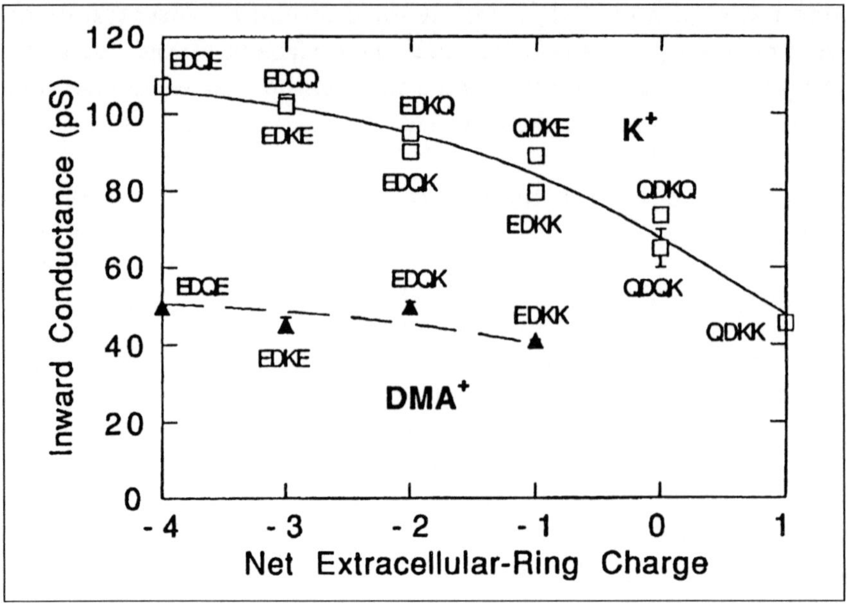

Fig. 6.3. Single channel conductance of the AChR as a function of the charge in the anionic rings. The lines are the model predictions. Reprinted with permission from Kienker P et al, Biophys J 1994; 66:325-34. © 1994, The Biophysical Society.

the intermediate ring, need not be in a single file. The surface charge model is appropriate to describe a situation in which the pore cannot be depleted of ions at low ionic strength.

The charges of the rings actively participate in ion transport. E. von Kitzing used a molecular model of the transmembrane segment to compute the energy of ions moving through the pore. Energy calculations indicated that the flexible side chains of glutamate and aspartate receive the ions at one end of the pore, move together with the ion, and pass them to the next residue. In this way ions can travel several Ångstroms with virtually no friction.[58]

Contribution of the Ring to Ion Selectivity

If the role of the rings of charge in ion transport is not yet clear, their participation in ion selection is even more obscure. Konno spent several years trying to understand how the ring charges determine ion selectivity in the AChR.[42] He estimated ion selectivity by conductance and by reversal potentials of wild type and mutant *Torpedo* AChR. Among the several mutant channels analyzed, the best studied was the one carrying the δE255Q mutant, in which one of the negative charges in the intermediate ring was neutralized. The decrease in conductance produced by the equivalent mutation in the α subunit (αE241Q) was so large it made it impractical to study. Both conductance and permeability ratio decreased, mostly for large ions in the mutant δE255Q. This was not an artifact of the normalization procedure, because K^+, which has an intermediate ion size, was used as a reference. The effect of the mutation δE255Q can be quantitatively described assuming an increase in the central barrier for Cs^+ in the energy profile. The result brought two surprises. Firstly, the large ions that are normally less hydrated were the most affected, contrary to Eisenman's selectivity theory in which a decrease in field strength will hinder the dehydration of small ions.[59] Secondly, not the wells, but the barriers were affected. If the rings were to provide a binding site for cations one would expect them to affect the minima in the energy profile.

Anion Permeability

The AChR is a classical example of a cation selective channel. Takeuchi et al[60] demonstrated that the reversal potential of endplate currents is independent of the external Cl^- concentration (Fig. 6.4A). Furthermore, Cl^- can be replaced by nitrate, sulfate, or glutamate with no change in the reversal potential. They concluded that the endplate channels are impermeable to anions.

Quantitative determinations of anion permeability, performed in rat cardiac ganglia, which probably express neuronal nicotinic AChR, indicated that the permeability to Cl^- (P_{Cl}/P_{Na}) is smaller than 0.005.[61] From what we know about the pore structure (see also chapter 5 in this volume), we assume that the cluster of negative charges or anionic rings is responsible for the cationic selectivity. This negative charge would repel anions from the entrances of the pore. Once the ions are inside the pore, they will be conducted according to their size. In agreement with this notion, electrostatic calculations on a model of the AChR showed that an hydroxyl ring can select either anions or cations. GABA receptors are anion selective channels that also contain hydroxyl residues in the M2 segment, which probably forms the wall of the pore, as is the case with the AChR. The hydroxyl group of serine and threonine side chains can act as ambidextrous ligands for cations or anions according to the orientation of the proton. On the one hand, when a cation is in the

Fig. 6.4. The AChR is impermeable to Cl. (A) The reversal potential of synaptic currents is independent of the external chloride concentration. Redrawn from Takeuchi A, Takeuchi N. On the permeability of the end-plate membrane during the action of transmitter. Reprinted with permission from Takeuchi A et al, J Physiol 1960; 154: 52-67. ©1960, Cambridge University Press. (B) Reversal potential as a function of the Cl concentration of the α7-1 (α7-P236(+)*E237A*S240G*V251T*L254S*L255G*E258N) mutant. Modified from Galzi JL et al, Nature 1992; 359 (6395):500-5. ©1992, Macmillan Journals Ltd.

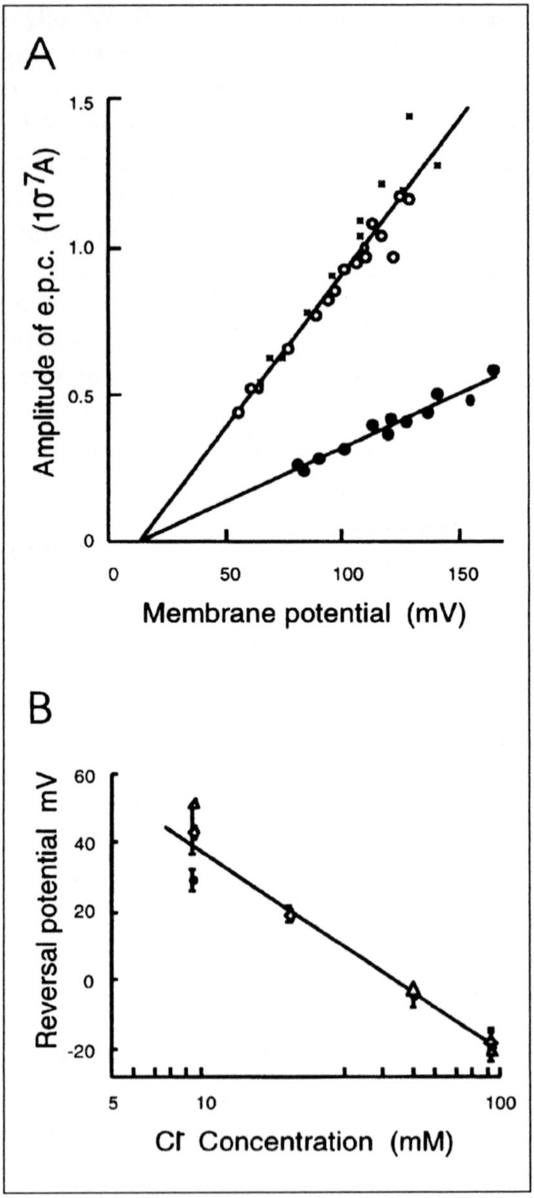

pore the proton will point away, making the cation interact with the lone electron pairs of the oxygen. On the other hand, when an anion is placed in the pore, the protons will be pointing in, interacting with the anion. Electrostatic calculations in a model of the selectivity filter indicated that hydroxyl filters are able to generate nonmonotonic selectivity sequences, following the Eisenman II sequence for cations (Rb>Cs>K>Na>Li) and the sequence AIX (Cl>Br>F>I) for anions.[62] Exactly the same sequence has been found for permeabilities measured from reversal potentials in the fetal type AChR.[63] Thus the AChR pore wall per se can allow both anions and cations to go through; it is the charges of the rings that determine the anion versus cation selectivity.

If the anion selectivity is determined by the rings of negative charges it would be possible to convert the AChR into an anion selective channel just by placing positive instead of negative charges at the anionic rings. Changeux and his colleagues explored this possibility using the homomeric neuronal AChR formed by the $\alpha 7$ subunit.[64] After comparing the amino acid sequence of the M2 segment of the $\alpha 7$ subunit with that of the GABA and glycine receptors (both anion selective channels, see chapter 2 of this volume), they noticed that there were only a few differences. The main striking feature of the GABA receptor subunits is the positive instead of negative charges at the extremes of the M2 segment. The internal ring contains an aspartate (D) in all but GABA-$\alpha 1$ receptors. The intermediate ring is replaced by an alanine and an extra proline or alanine is inserted in the previous position. The extracellular ring has either an alanine (A) in the case of glycine receptors, or an asparagine and aspartate in the case of $\alpha 1$ and $\alpha 2$ GABA receptors respectively.

Mutation of only three amino acids were sufficient to obtain an anion selective channel: the αE237A replacement in the intermediate ring, the αV251T replacement in the middle of the M2 segment, and insertion of a proline before position 237. The resulting channel had a perfect chloride selectivity, as indicated by reversal potential measurements in different extracellular Cl^- concentrations, which followed the Nernst relationship (see Fig. 6.4B and Eq. 4).

The experiments with the $\alpha 7$ AChR indicated that in order to achieve anion permeability, reverting the sign of the anionic ring charges is not sufficient. The process required the insertion of a proline, suggesting an alteration in tertiary structure. This suggests the possibility that a different surface of the M2 segment faces the pore in an anion selective channel. In agreement with this idea is the αV251T requirement which will make that side of the (putative) α-helix more polar such that the ambidextrous threonine faces the lumen of the pore.

The anion/cation selectivity of ligand-gated channels is still an open problem that will certainly be the topic of future research. The question whether charges are the only structural elements responsible for the anion versus cation selectivity remains unanswered. Could it be possible to find a ligand-gated channel which switches selectivity according to the charges of the rings? The experiments of Changeux and coworkers provided some clues. Perhaps trying to convert a GABA receptor into a cation channel could be another approach. The anion selectivity of the AChR mutated in the intermediate ring should also be explored. The potential anion selectivity of the mutant αE261K (using the numbers of the rat AChR) and equivalent residues at the other subunits need to be examined. Observations made in the glutamate receptor formed by the GluR-6(R) where selectivity is switched from anionic to cationic just by a single amino acid substitution[65] could also apply to the AChR. The channel formed by GluR-6(Q) is almost perfectly cation selective, whereas that formed by GluR-6(R) has $P_{Cl}/P_{Cs} = 0.74$. The difference between these two homomeric receptors is the neutral glutamine (Q) and the positively charged asparagine (R) in the narrow region of the pore.

A second aspect that needs to be explored is the mechanism by which the rings make a channel cation selective. One possibility is that the rings act by a purely electrostatic effect, in which case the cation selectivity would vanish at high ionic strength. Another possibility is that the rings participate in ion binding and in the translocation of ion through the pore. In this case differences would be observed between rings made of aspartate and glutamate, which differ in their side chain length and flexibility.

Divalent Cations

The AChR is permeable to Ca^{2+} and other divalent ions. At the frog neuromuscular junction, Ca^{2+} influx was demonstrated using the Ca^{2+}-sensitive dye Arsenazo. The Ca^{2+} influx varied exponentially with the membrane potential, and linearly with the external Ca^{2+} concentration. The magnitude was up to 10% of the total current.[66]

A quantitative characterization of the divalent ions influx through the AChR required a more uniform control of potential in the muscle fiber preparation, as well as a control of the internal solution. Hille and his group used the Vaseline-gap technique to determine the permeability of the frog neuromuscular junction to Ca^{2+} and other divalent ions.[40] Among the divalent ions, the most permeant was Mg^{2+} ($P_{Mg}/P_{Na} = 0.25$). The permeability ratio for other divalent ions followed the sequence $Ca^{2+}>Ba^{2+}>Sr^{2+}$. Four transition metals were also permeant. They had permeability ratios (P_x/P_{Na}) similar to that of Mg^{2+}. Cd^+ had a low permeability ($P_{Cd}/P_{Na} = 0.13$) that was attributed to the formation of complexes with Cl^- which may reduce the free Cd^+ concentration. The selectivity sequence for $Mg^{2+}>Ca^{2+}>Ba^{2+}>Sr^{2+}$ was inversely related to their mobility in water, as if the ions traverse the pore without friction against the pore wall. This interpretation is consistent with our present knowledge of the pore size of the AChR. The hydrated radii for these series of divalent ions varies from 4.28 Å (Mg^{2+}) to 4.04 Å (Ba^{2+}).[67] Divalent ions are likely to move through the AChR channel together with their complete hydration shell.

Ca^{2+} Permeability

Early studies in rat myotubes showed that the reversal potential was insensitive to external Ca^{2+} concentrations up to 30 mM,[68] suggesting that the AChR is impermeable to Ca^{2+}. Determinations on eel electroplax, on the other hand, suggested a P_{Ca}/P_K of 0.7.[69] Flux measurements in culture-muscle cells yielded a P_{Ca}/P_{Na} of 0.2.[13] The Ca^{2+} permeability (P_{Ca}/P_{Na}) of the frog endplate receptors was 0.16 and 0.29 at 114 and 2.5 mM Ca respectively.[40] Dani and his group determined the Ca^{2+} permeability of the AChR from the mouse-derived clonal cell line BC3H-1, a cell line that expresses mainly the $\alpha_2\beta\gamma\delta$ form of the receptor.[70] The permeability for Ca^{2+} measured in 20 mM Ca^{2+} ($P_{Ca}/P_{Na} = 0.2$)[71] was comparable to that measured for the frog neuromuscular junction at a similar Ca^{2+} concentration ($P_{Ca}/P_{Na} = 0.22$).[40] The Ca^{2+} permeability of native AChR from dissociated muscle, measured in 100 mM Ca^{2+}, was 0.1 and 0.2 for fetal and adult muscle, respectively.[72] The recombinant AChR from rat expressed in *Xenopus* oocytes reproduced these values. The reversal potential for Ca^{2+} of the $\alpha_2\beta\gamma\delta$ form of the AChR was similar to that measured in muscle dissociated from neonatal rat, and that of the $\alpha_2\beta\epsilon\delta$ form was similar to that measured in muscle dissociated from adult rat.[72]

Permeability Is Higher at Low Ca^{2+}

The Ca^{2+} permeability (P_{Ca}/P_{Na}) of AChR from frog endplate showed dependence on the external Ca^{2+} concentration, varying from 0.29 for 5 mM Ca^{2+} to 0.16 for 114 mM Ca^{2+} concentration.[40] The Ca^{2+} permeability of native AChR from neonatal (P9) rat endplate also varied with Ca^{2+} concentration. It was higher at low Ca^{2+} concentrations. At physiological Ca^{2+} levels (2 mM), $P_{Ca}/P_{Cs} = 1$ for the fetal type, which is formed of the $\alpha\beta\gamma\delta$ subunits. The permeability of the adult type native receptor was even higher. At 2 mM Ca^{2+} $P_{Ca}/P_{Cs} = 2$ (Fig. 6.5). Upon increasing the Ca^{2+} concentration, the permeability for Ca^{2+} decreased to 0.1 and 0.2 for the fetal and adult native AChR, respectively.[72] This change in Ca^{2+} permeability with concentration presumably reflects the effect of fixed charges at the surface of the channel pore that alter the local concentration of Ca^{2+}.

Ca^{2+} Conductance

Bregestovski and his colleagues determined single-channel conductance for Ca^{2+} in the frog neuromuscular junction using noise analysis.[73] At depolarizing potentials in a solution containing 160 mM Ca^{2+}, the single-channel conductance for Ca^{2+} was about 6 pS for hyperpolarizing potentials. In Ringer's solution the conductance of the endplate channels was 30 pS. Conductance for Ca^{2+} was measured directly in BC3H-1 cells using single-channel recording. In pure 110 mM Ca^{2+} the conductance at negative potentials was 12 pS.[74] For comparison, in 150 mM Na^+ the conductance was 51 pS. Using a barrier model, Dani and his group estimated that 2% of the inward current at negative potential would be carried by Ca^{2+} in physiological conditions. Quantitative measurements of Ca^{2+} flux using microfluorimetry confirmed these results.[75] The recombinant rat AChR in 100 mM external Ca^{2+} has conductances of 15 and 25 pS in its fetal and adult forms, respectively (Villarroel, unpublished observations).

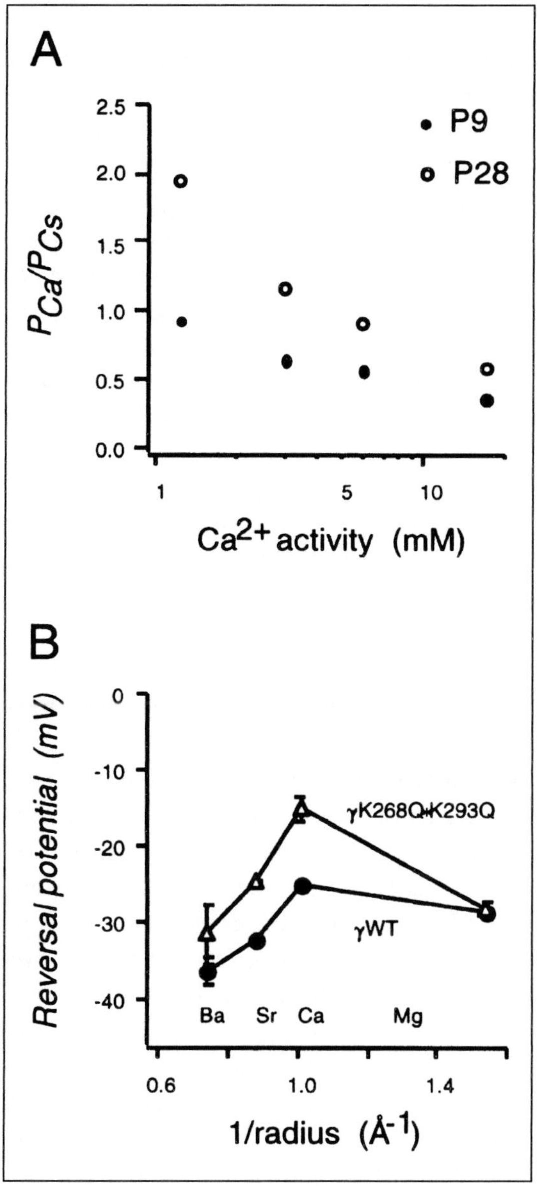

Fig. 6.5. Conduction of divalent ions. (A) Permeability of the AChR at low Ca^{2+} concentration. Reprinted with permission from Villarroel A et al, J Physiol 1996; 496 (2):331-338. © 1996, The Physiological Society. (B) Selectivity sequences for divalent ions of the wild type and γK268Q*K293Q AChR. Notice that Mg permeability is not affected by the mutation (Villarroel and Sakmann unpublished).

Neuronal Receptors

Ca^{2+} ion permeation has been studied not only in receptors from muscle, or muscle-related tissue, but also in the neuronal nicotinic AChRs. These receptors present in the brain have a higher Ca^{2+} permeability than that of the muscle AChR.[75] The Ca^{2+} permeability of neuronal AChR from rat sympathetic ganglion was $P_{Ca}/P_{Na} = 0.65$.[76] The neuronal receptor expressed in PC12 cells has higher Ca-permeability ($P_{Ca}/P_{Na} = 1.75$).[77] The native α-bungarotoxin sensitive receptor from hippocampus has a P_{Ca}/P_{Cs} of 6.1 which is unusual among neuronal AChRs.[78] An unusually high Ca^{2+} permeability has only been found in the α7-type AChR, indicating that the hippocampal receptor is probably identical to α7. The Ca-permeability in α7 AChR is extremely sensitive to mutations in the pore. Almost any mutation in the M2 segment will abolish the Ca^{2+}-permeability.[79] The single-channel conductance of the nicotinic AChR from the *nucleus habenuale medialis* was 10 pS in 100 mM Ca^{2+}.[80]

P_f vs Reversal Potentials

The classical way to estimate efficacy in ion permeation is to determine the permeability ratios from reversal potentials. The permeability ratio will give the ratio of the fluxes at the reversal potential. In the case of the Ca^{2+} permeation it is possible to visualize and to quantify the Ca^{2+} influx using fluorescence microfluorimetry.[81,82] Zhou and Neher,[4] using the Ca-sensitive dye Fura-2, determined the fractional Ca^{2+} current in chromaffin cells. When Fura-2 is present in a concentration of 1 mM or higher, it overcomes the weak endogenous Ca^{2+} buffers in their ability to bind Ca^{2+}. Since all Ca^{2+} ions that enter the cell are trapped by Fura-2, the change of the Ca-sensitive fluorescence signal is proportional to Ca influx. In this way the pure Ca^{2+} influx can be determined by the Fura-2 signal.

Pf is defined as the fraction of the total current that is carried by Ca^{2+}. The influx of Ca^{2+} leads to a Ca^{2+} accumulation inside the cell. Therefore the total charge accumulated, which is determined as the integral of the current, is proportional to the increase in the fluorescence signal. In order to determine the proportionality constant, a calibration measurement is done where a known fraction of the current is carried by Ca^{2+} (i.e., the voltage-gated Ca^{2+} current). The method consists of simultaneous measurements of fluorescence intensity and ionic current. P_f is determined as the ratio between the fluorescent intensity and the maximum fluorescence. They found that in chromaffin cells, in 2 mM Ca^{2+} and at -70 mV, Ca^{2+} contributes 2.5% of the net current.[13] In sympathetic ganglia the influx of Ca^{2+} through voltage-gated calcium channels was 46 times that obtained for neuronal AChR in the same preparation, indicating a P_f of 2.2%.[83] Dani introduced a different calibration procedure to obtain the maximal measurements by using a medium with Ca^{2+} as the only permeant cation.[84] Using this procedure the muscle-type AChR from BC3H-1 cells in 2.5 mM Ca^{2+} and at -50 mV had a P_f of 2.0%. Under similar conditions, AChR from chromaffin cells carried 4.1% of the total current.[75]

Using the Goldman-Hodgkin-Katz formalism, the fractional Ca^{2+} current at a given potential can be related to the permeability ratio (P_{Ca}/P_{Na}) as:[82,85]

$$P_f = \frac{1}{1+\left(\frac{1}{4}\frac{P_{Ca}}{P_{Na}}\frac{[Na]}{[Ca]_o}\left(1-\exp\left(\frac{2FV}{RT}\right)\right)\right)} \tag{8}$$

Notice that the permeability ratio P_{Ca}/P_{Na} which is taken as a constant, may be concentration and voltage dependent. Experimentally, the fractional-calcium current is voltage dependent. The neuronal AChR has a $P_f = 2.7$ at -10 mV, compared to 4.1 at -50 mV.[75] Near the reversal potential, as the total current decreases, the P_f diverges. In addition, P_f is also dependent on the Ca^{2+} concentration. The muscle-type receptor from BC3H-1 has a P_f of 3.4 at 5 mM Ca^{2+}. The neuronal AChR, which is more permeable to Ca^{2+}, has a P_f of 6.3 under similar conditions.[75]

Charges Change Ca^{2+}-Permeability

Even though we do not know the exact factors that make a pore Cα2+ selective, we can speculate on the basis of permeability measurements of either close-related, or point mutated, channels. There are two studies that provide some clues. In the muscle AChR the adult form (α2βεδ) has larger Ca^{2+} permeability than that of the fetal form ($\alpha_2\beta\gamma\delta$);[72] therefore, one may infer that amino acid residues which differ between the γ- and ε-subunits may be responsible for the differences in Ca^{2+} permeability. The obvious difference in the amino acid sequence is a pair of charged residues at both ends of the M2 segment. These residues were found to be important determinants of the conductance differences between the fetal and adult channel.[44] Since Ca^{2+} is a divalent ion, it is likely to interact strongly with charges along the conduction pathway.

As expected, mutation γK268Q increases the permeability of the AChR.[86] In this mutated channel the permeability for other monovalent ions also increases. The double mutation γK268Q*K293Q produces a larger increase in divalent ion permeability (Fig. 6.5B). From such observations we learn that charges participate in the permeation of divalent ions. The removal of two positive charges favors the conduction of Ca^{2+} by a mechanism which may involve an increase in local concentration produced by fixed charges. In a drift mechanism framework, the removal of positive charges may increase the equilibrium concentration of Ca^{2+} inside the channel.

Perspectives

In this section I will summarize what we know and do not know about ion conduction in the AChR.

The AChR is a cation-selective channel. This fact is well established. The cationic selectivity is determined in some way by the presence of the anionic rings. Among monovalent cations, when selectivity is measured by reversal potentials, Cs^+ is preferred. In mammalian AChRs, however, Rb^+ is preferred. The ion exhibiting the highest conductance is, however, K^+. We understand this as the restricted ion flow of large cations by the selectivity filter, which selects ions according to size. As a consequence, ions that cross the AChR pore are only partially dehydrated. The AChR is permeable to Ca^{2+}. In the case of rat receptor, the Ca-permeability is high. In particular, the adult form of the rat AChR has a P_{Ca}/P_{Cs} of 2 when measured at physiological Ca^{2+} levels. The anionic rings contribute to Ca-permeability, as the difference in charges in the M2 segments of the $\alpha_2\beta\gamma\delta$ and $\alpha_2\beta\varepsilon\delta$ receptor suggest.

The distinction between peripheral (muscle and *Torpedo*) and central (neuronal) AChR is now becoming clearer. The muscle receptor has a Ca-permeability 10 times lower[83] than that of the neuronal AChR.[87] The permeability selectivity sequence of neuronal AChR is Cs<Li<Na<K<Ca.[83] The narrow region of the pore is probably located at a different position in these two receptor families.[31] In particular, the narrow region of the muscle and *Torpedo* AChR is located more towards the intracellular side.[31] Surprisingly, there is no data on the pore size of the neuronal AChR. Based on the similari-

ties between the latter and the serotonin (5-HT_3) receptor, the pore size of the neuronal AChR is probably larger than that of both muscle and *Torpedo* AChRs. The extrapolation of findings from the muscle and *Torpedo* AChR to neuronal receptor now seems as inappropriate as the comparison between AMPA/kainate and NMDA receptors, which in spite of sharing the same neurotransmitter, have completely different conduction properties.[88]

If we were to write a list of the points that we ignore, it would probably be as large as this chapter. The following are some of the current problems in AChR ion conduction:

There is a clear discrepancy in the alkali-cation selectivity of the AChR when measured by conductance and by reversal potentials. Traditionally, permeability ratio measures the maximum in the energy profile, whereas conductance is also sensitive to the wells. Point mutations in the M2 segment produce changes both in conductance and permeability ratio. One may infer then that a mutation that changes the pore size would modify the peak, leaving the well untouched. Surprisingly, no mutation of this type has produced a significant change in the shape of the current-voltage curves, as would be expected if a particular location in the energy profile is changed. Charge mutations also produce equivalent results.

A second aspect that will be a topic of future research is the mechanism of Ca^{2+} permeability. So far there is only an inventory of mutations that change Ca-permeability, but a physical picture is missing. We would like to understand Ca-permeability in terms of the properties of side chain residues in the pore. A combination of electrophysiology and optical methods certainly will add new insight into Ca^{2+} conduction.

Anionic rings have been known for almost a decade.[54] They are important because they increase the conductance of the AChR. The mechanism by which they operate, however, is still unknown. The hypothesis that the rings act by an electrostatic mechanism, even though attractive, is completely inconsistent with the finding that the conductance changes are independent of ionic strength.[56] A new permeation model in which the effects of the charges are built into the equation that describes the conductance-concentration is expected to be more promising than the amended Michaelis-Menten model.[89]

The pore size of the AChR is 7.4Å when determined by permeation of organic ions, and 9-10 Å as deduced from electron microscope images.[32] Part of the difference is probably due to a layer of water molecules which adheres to the surface of the pore. It has now become clear that the first hydration layer of proteins has different properties than that of the other layers.[33] The hypothesis of a layer of water also solves another discrepancy on whether size[20] or hydrophobicity[16] of the side-chain residue in the selectivity filter is *the* determinant of ion flow. An increase in hydrophobicity would be equivalent to an increase in size of the hydrophobic side chain, and is likely to alter the first layer of water.

The beginning of a new era in AChR research will begin when we comprehend the structure of the protein at a resolution high enough to be able to discern individual atoms. Then molecular models will become more realistic, and we will be able to accompany the ion to each place visited during its journey through the pore.

References

1. Neher E, Sakmann B. Single-channel currents recorded from membrane of denervated frog muscle fibers. Nature 1976; 260(5554): 799-802.
2. Noda M, Takahashi H, Tanabe T et al. Primary structure of α-subunit precursor of Torpedo californica acetylcholine receptor deduced from cDNA sequence. Nature 1982; 299(5886):793-7.
3. Noda M, Takahashi H, Tanabe T et al. Structural homology of Torpedo californica acetylcholine receptor subunits. Nature 1983; 302(5908):528-32.
4. Miledi R, Parker I, Schalow G. Transmitter induced calcium entry across the postsynaptic membrane at frog end-plates measured using arsenazo III. J Physiol 1980; 300(197):197-212.
5. Methfessel C, Witzemann V, Takahashi T et al. Patch clamp measurements on *Xenopus laevis* oocytes: currents through endogenous channels and implanted acetylcholine receptor and sodium channels. Pflügers Arch 1986; 407(6):577-88.
6. Zhou Z, Neher E. Calcium permeability of nicotinic acetylcholine receptor channels in bovine adrenal chromaffin cells. Pflügers-Arch 1993; 425(5-6):511-7.
7. Mishina M, Takai T, Imoto K et al. Molecular distinction between fetal and adult forms of muscle acetylcholine receptor. Nature 1986; 321(6068):406-11.
8. Gu Y, Hall ZW. Immunological evidence for a change in subunits of the acetylcholine receptor in developing and denervated rat muscle. Neuron 1988; 1(2):117-25.

9. Schoepfer R, Whiting P, Esch F et al. cDNA clones coding for the structural subunit of a chicken brain nicotinic acetylcholine receptor. Neuron 1988; 1(3):241-8.
10. Nef P, Oneyser C, Alliod C et al. Genes expressed in the brain define three distinct neuronal nicotinic acetylcholine receptors. Embo J 1988; 7(3):595-601.
11. Edward JT. Molecular volumes and the Stokes-Einstein equation. J Chem Ed 1970; 47(4):261-270.
12. Richards FM. The interpretation of protein structures: Total volume, group volume distributions and packing density. J Mol Biol 1974; 82:1-14.
13. Huang LY, Catterall WA, Ehrenstein G. Selectivity of cations and nonelectrolytes for acetylcholine-activated channels in cultured muscle cells. J Gen Physiol 1978; 71(4):397-410.
14. Dwyer TM, Adams DJ, Hille B. The permeability of the endplate channel to organic cations in frog muscle. J Gen Physiol 1980; 75(5):469-92.
15. Sanchez JA, Dani JA, Siemen D et al. Slow permeation of organic cations in acetylcholine receptor channels. J Gen Physiol 1986; 87(6):985-1001.
16. Cohen BN, Labarca C, Davidson N et al. Mutations in M2 alter the selectivity of the mouse nicotinic acetylcholine receptor for organic and alkali metal cations. J Gen Physiol 1992; 100(3):373-400.
17. Villarroel A, Sakmann B. Threonine in the selectivity filter of the acetylcholine receptor channel. Biophys J 1992; 62(1):196-205.
18. Pimentel GC, McClellan AL. The Hydrogen Bond. San Francisco and London: Freeman and Co., 1960.
19. Sandorfy C. Anarmonicity and hydrogen bonding. In: The Hydrogen Bond: Recent Developments in Theory and Experiments. Schuster P, Zundel G, Sandorfy C, eds. Amsterdam: North-Holland, 1976.
20. Villarroel A, Herlitze S, Koenen M et al. Location of a threonine residue in the α-subunit M2 transmembrane segment that determines the ion flow through the acetylcholine receptor channel. Proc R Soc Lond B Biol Sci 1991; 243(1306):69-74.
21. Cohen BN, Labarca C, Czyzyk L et al. $Tris^+/Na^+$ permeability ratios of nicotinic acetylcholine receptors are reduced by mutations near the intracellular end of the M2 region. J Gen Physiol 1992; 99(4):545-72.
22. Oiki S, Madison V, Montal M. Bundles of amphipathic transmembrane alpha-helices as a structural motif for ion-conducting channel proteins: studies on sodium channels and acetylcholine receptors. Proteins 1990; 8(3):226-36.
23. Wang F, Imoto K. Pore size and negative charge as structural determinants of permeability in the *Torpedo* nicotinic acetylcholine receptor channel. Proc R Soc Lond B Biol Sci 1992; 250(1327):11-7.

24. Rosenberg PA, Finkelstein A. Interaction of ions and water in gramicidin A channels: streaming potentials across lipid bilayer membranes. J Gen Physiol 1978; 72(3):327-40.
25. Cecchi X, Bull R, Franzoy R et al. Probing the pore size of the hemocyanin channel. Biochim Biophys Acta 1982; 693(1):173-6.
26. Dani JA. Open channel structure and ion binding sites of the nicotinic acetylcholine receptor channel. J Neurosci 1989; 9(3):884-92.
27. Charnet P, Labarca C, Leonard RJ et al. An open-channel blocker interacts with adjacent turns of α-helices in the nicotinic acetylcholine receptor. Neuron 1990; 4(1):87-95.
28. Neher E, Steinbach JH. Local anaesthetics transiently block currents through single acetylcholine-receptor channels. J Physiol Lond 1978; 277:153-76.
29. Skok VI, Groisman SD, Melnitchenko LV et al. Selective pharmacological blockade of parasympathetic and enteric ganglia. J Auton Nerv Syst 1991; 35(3):211-7.
30. Skok VI, Voitenko SV, Kurenniy DE et al. The ionic channel of neural nicotinic acetylcholine receptors is funnel-shaped. Neuroscience 1995; 67(4):933-9.
31. Sankararamakrishnan R, Adcock C, Sansom MSP. The pore domain of the nicotinic acetylcholine receptor: Molecular modelling, pore dimensions, and electrostatics. Biophys J 1996; 71:1659-1671.
32. Unwin N. Acetylcholine receptor channel imaged in the open state. Nature 1995; 373(6509):37-43.
33. Marsh D. Peptide models for membrane channels. Biochem J 1996; 315:345-361.
34. Otting G, Liepinsh E, Wuthrich K. Protein hydration in aqueous solution. Science 1991; 254(5034):974-80.
35. Overton E. Beiträge zur allgemeinen Muskel- und Nervenphysiologie. II. Über die Unentbehrlichkeit von Natrium- (oder Lithium-) Ionen für den Kontraktionsakt des Muskels. Pflügers Archiv 1902; 92: 346-386.
36. Fatt P. The electromotive action of acetylcholine at the motor endplate. J Physiol 1950; 111:408-422.
37. Fatt P, Katz B. An analysis of the end-plate potential recorded with an intracellular electrode. J Physiol 1951; 115:320-370.
38. Takeuchi A, Takeuchi N. On the permeability of the end-plate membrane during the action of transmitter. J Physiol 1960; 154:52-67.
39. Eisenman G, Horn R. Ionic selectivity revisited: the role of kinetic and equilibrium processes in ion permeation through channels. J Membr Biol 1983; 76(3):197-225.
40. Adams DJ, Dwyer TM, Hille B. The permeability of endplate channels to monovalent and divalent metal cations. J Gen Physiol 1980; 75(5):493-510.
41. Eisenman G, Krasne SJ. The ion selectivity of carrier molecules, membranes and enzymes. In: MTP International Review of Science, Biochemistry Series. Fox CE, ed. 1975; 2:27-59, London: Butterworths.

42. Konno T, Busch C, Von K-E et al. Rings of anionic amino acids as structural determinants of ion selectivity in the acetylcholine receptor channel. Proc R Soc Lond Biol 1991; 244(1310):69-79.
43. Imoto K, Methfessel C, Sakmann B et al. Location of a δ-subunit region determining ion transport through the acetylcholine receptor channel. Nature 1986; 324(6098):670-674.
44. Herlitze S, Villarroel A, Witzemann V et al. Structural determinants of channel conductance in fetal and adult rat muscle acetylcholine receptors. J Physiol Lond 1996; 492(3):775-787.
45. Witzemann V, Barg B, Nishikawa Y et al. Differential regulation of muscle acetylcholine receptor γ- and ε-subunit mRNAs. FEBS Lett 1987; 223(1):104-12.
46. Quartararo N, Barry PH, Gage PW. Ion permeation through single channels activated by acetylcholine in denervated toad sartorius skeletal muscle fibers: effects of alkali cations. J Membr Biol 1987; 97(2):137-59.
47. Leonard RJ, Labarca CG, Charnet P et al. Evidence that the M2 membrane-spanning region lines the ion channel pore of the nicotinic receptor. Science 1988; 242(4885):1578-81.
48. Imoto K, Konno T, Nakai J et al. A ring of uncharged polar amino acids as a component of channel constriction in the nicotinic acetylcholine receptor. FEBS Lett 1991; 298(2):193-200.
49. Otting G, Liepinsh E, Wüthrich K. Protein hydration in aqueous solution. Science 1991; 254(5034):974-80.
50. Villarroel A, Herlitze S, Witzemann V et al. Asymmetry of the rat acetylcholine receptor subunits in the narrow region of the pore. Proc R Soc Lond B Biol Sci 1992; 249(1326):317-24.
51. Yu L, Leonard RJ, Davidson N et al. Single-channel properties of mouse-Torpedo acetylcholine receptor hybrids expressed in *Xenopus* oocytes. Brain Res Mol Brain Res 1991; 10(3):203-11.
52. Bouzat C, Bren N, Sine SM. Structural basis of the different gating kinetics of fetal and adult acetylcholine receptors. Neuron 1994; 13(6):1395-402.
53. Dani J. Ion-channel entrances influence permeation. Net charge, size, shape, and binding considerations. Biophys J 1986; 49:607-618.
54. Imoto K, Busch C, Sakmann B et al. Rings of negatively charged amino acids determine the acetylcholine receptor channel conductance. Nature 1988; 335(6191):645-8.
55. Grahame D. The electric double layer and the theory of electrocapilarity. Chem Rev 1974; 41:441-501.
56. Kienker P, Tomaselli G, Jurman M et al. Conductance mutations of the nicotinic acetylcholine receptor do not act by a simple electrostatic mechanism. Biophys J 1994; 66:325-34.
57. Palma A, Li L, Chen XJ et al. Effects of pH on acetylcholine receptor function. J Membr Biol 1991; 120(1):67-73.

58. von Kitzing E. Structure modeling of the Acetylcholine receptor channel and related ligand gated channels. In: Kluwer and Dortrecht, eds. Modelling of Biomolecular Structures and Mechanisms. Academic Publishers, 39-57.
59. Eisenman G. Cation selective glass electrodes and their mode of operation. Biophys J 1962; (2):259-323.
60. Takeuchi A, Takeuchi N. On the permeability of the end-plate membrane during the action of transmitter. J Physiol 1960; 154:52-67.
61. Fieber LA, Adams DJ. Acetylcholine-evoked currents in cultured neurones dissociated from rat parasympathetic cardiac ganglia. J Physiol 1991; 434:215-37.
62. Eisenman G, Villarroel A. Ion selectivity of pentameric protein channels: Backbone carbonyl ligands as cation binding ligands and side chain hydroxyls as "ambidextrous" ligands for cations and anions in viral capsids. In: Pasternak CA, ed. Monovalent Cations in Biological Systems. Boca Raton, Florida: CRC Press, 1990.
63. Villarroel A, Herlitze S, Witzemann V et al. Asymmetry of the rat acetylcholine receptor subunits in the narrow region of the pore. Proc R Soc Lond B Biol Sci 1992; 249(1326):317-24.
64. Galzi JL, Devillers-Thiéry A, Hussy N et al. Mutations in the channel domain of a neuronal nicotinic receptor convert ion selectivity from cationic to anionic. Nature 1992; 359(6395):500-5.
65. Burnashev N, Villarroel A, Sakmann B. Dimensions and ion selectivity of recombinant AMPA and kainate receptor channels and their dependence on Q/R site residues. J Physiol 1996; 496(1):165-73.
66. Miledi R, Parker I, Schalow G. Transmitter induced calcium entry across the postsynaptic membrane at frog end-plates measured using arsenazo III. J Physiol Lond 1980; 300(197):197-212.
67. Nightingale ER. Phenomenological theory of ion solvation. Effective radii of hydrated ions. J Phys Chem 1959; 63:1381-1387.
68. Ritchie AK, Fambrough DM. Ionic properties of the acetylcholine receptor in cultured rat myotubes. J Gen Physiol 1975; 65(6):751-67.
69. Lassignal NL, Martin AR. Effect of acetylcholine on postjunctional membrane permeability in eel electroplaque. J Gen Physiol 1977; 70(1):23-36.
70. Criado M, Koenen M, Sakmann B. Assembly of an adult type acetylcholine receptor in a mouse cell line transfected with rat muscle ε-subunit DNA. FEBS Lett 1990; 270(1-2):95-9.
71. Vernino S, Amador M, Luetje CW et al. Calcium modulation and high calcium permeability of neuronal nicotinic acetylcholine receptors. Neuron 1992; 8(1):127-34.
72. Villarroel A, Sakmann B. Calcium permeability increase of end-plate channels in rat muscle during postnatal development. J Physiol 1996; 496(2):331-338.

73. Bregestovski PD, Miledi R, Parker I. Calcium conductance of acetylcholine-induced endplate channels. Nature 1979; 279(5714):638-9.
74. Decker ER, Dani JA. Calcium permeability of the nicotinic acetylcholine receptor: The single-channel calcium influx is significant. J Neurosci 1990; 10(10):3413-20.
75. Vernino S, Rogers M, Radcliffe KA et al. Quantitative measurement of calcium flux through muscle and neuronal nicotinic acetylcholine receptors. J Neurosci 1994; 14(9):5514-24.
76. Adams DJ, Nutter TJ. Calcium permeability and modulation of nicotinic acetylcholine receptor-channels in rat parasympathetic neurons. J Physiol Paris 1992; 86(1-3):67-76.
77. Sands SB, Barish ME. Calcium permeability of neuronal nicotinic acetylcholine receptor channels in PC12 cells. Brain Res 1991; 560(1-2):38-42.
78. Castro NG, Albuquerque EX. α-Bungarotoxin-sensitive hippocampal nicotinic receptor channel has a high calcium permeability. Biophys J 1995; 68(2):516-24.
79. Bertrand D, Galzi JL, Devillers-Thiéry A et al. Mutations at two distinct sites within the channel domain M2 alter calcium permeability of neuronal α7 nicotinic receptor. PNAS 1993; 90(15):6971-5.
80. Mulle C, Choquet D, Korn H et al. Calcium influx through nicotinic receptor in rat central neurons: its relevance to cellular regulation. Neuron 1992; 8(1):135-43.
81. Neher E, Augustine GJ. Calcium gradients and buffers in bovine chromaffin cells. J Physiol Lond 1992; 450(273):273-301.
82. Schneggenburger R, Zhou Z, Konnerth A et al. Fractional contribution of calcium to the cation current through glutamate receptor channels. Neuron 1993; 11(1):133-43.
83. Trouslard J, Marsh SJ, Brown DA. Calcium entry through nicotinic receptor channels and calcium channels in cultured rat superior cervical ganglion cells. J Physiol Lond 1993; 468(53):53-71.
84. Rogers M, Dani J. Comparison of quantitative calcium influx through NMDA, ATP, and ACh receptor channels. Byophys J 1995; 68: 501-506.
85. Burnashev N, Zhou Z, Neher E et al. Fractional calcium currents through recombinant GluR channels of the NMDA, AMPA and kainate receptor subtypes. J Physiol London 1995; 485(2):403-418.
86. Villarroel A, Sakmann B. Distinct Ca^{2+} permeability of the adult and fetal acetylcholine receptor from rat muscle. Society for Neuroscience Abstracts 1995; 21:11.5.
87. Seguela P, Wadiche J, Dineley M-K et al. Molecular cloning, functional properties, and distribution of rat brain α7: a nicotinic cation channel highly permeable to calcium. J Neurosci 1993; 13(2):596-604.
88. Villarroel A. The pore of the NMDA receptor. Thai J Physiol 1995; 8(1):1-11.

89. Syganow A, von Kitzing E. The integral weak diffusion and diffusion approximations applied to ion transport through biological ion channels. J Phys Chem 1995; 99(31):12030-12040.

CHAPTER 7

Neuronal Nicotinic Acetylcholine Receptors‡

Ronald J. Lukas

Introduction

Nicotinic acetylcholine receptors (AChR) are prototypical members of the neurotransmitter-gated superfamily of ion channels (ionotropic neurotransmitter receptors; see Figure 7.1; see chapters 2, 5, and 6 in this volume).‡ AChR mediate some of the effects of the endogenous neurotransmitter, acetylcholine (ACh), and they are principal biological targets of the tobacco alkaloid, nicotine, which generally mimics actions of ACh at AChR.

"Neuronal" Nicotinic Receptor Subtypes

Genetic Basis for Subunit Diversity

Consistent with decades-old pharmacological findings and reinforced by new data obtained over the last 10 years or so, AChR exist as a variety of subtypes. At least part of this heterogeneity in AChR subtypes is derived from diversity in genes that encode AChR subunits. To date,‡ 16 AChR subunit genes have been cloned from

‡This chapter focuses on AChR found in mammalian/vertebrate neurons and principally cites recent reviews and a few, selected studies published from 1990[1] through 1996. It is inevitable and unintentional that many relevant reports, including those presenting initial discoveries in an area of study, will not be mentioned in this necessarily abbreviated overview. Bibliographies in the few research articles and reviews cited here provide a much more comprehensive literature concerning AChR in neurons.

The Nicotinic Acetylcholine Receptor: Current Views and Future Trends,
edited by Francisco J. Barrantes.

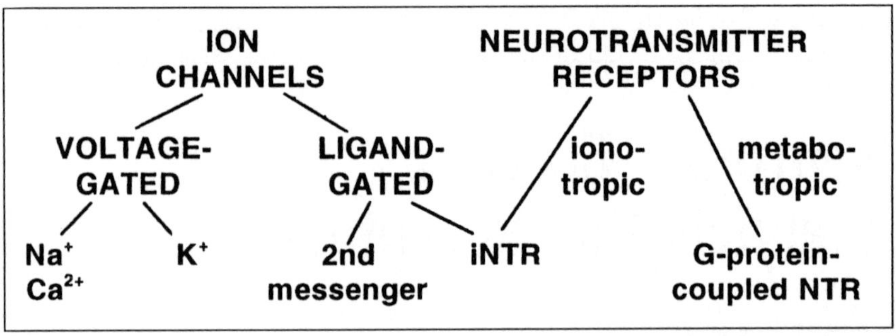

Fig. 7.1. Schematic "tree" diagram showing idealized relationships between "ion channels" and "neurotransmitter receptors" indicating placement of neurotransmitter-gated ion channels ("ionotropic" neurotransmitter receptors; "iNTR") at the junction between these two classes of signaling molecules. This classification focuses on these macromolecular groupings because of their defining roles in the nervous system in the mediation of cellular electrical excitability, electrical impulse propagation, receipt of extracellular chemical messages, and modulation of/via electrical and extracellular signaling by/of intracellular chemical messages. Included among the "voltage-gated" branch of ion channels are K^+ channels, whose subunits have six transmembrane domains and a hairpin loop-pore region (formerly called the H5 domain and now called the "P region") and assemble as tetramers, and Na^+ and Ca^{2+} channels, whose α subunits are built as four contiguous modules (pseudo-tetrameric), each with six K^+ channel subunit-like transmembrane domains and a P region. Included among the "ligand-gated" branch of ion channels, but distinct from ionotropic neurotransmitter receptors, are second messenger ("2nd messenger")-gated ion channels, such as the ryanodine (Ca^{2+}-activated), inositol triphosphate, and cyclic nucleotide receptors, all of which seem to have six transmembrane domains and a domain homologous to a P region per subunit. Perhaps there should be another bifurcation of the second messenger-gated branch of ion channels to include the inward rectifier superfamily of K^+ channels, opening of which is modulated by agents such as ATP and polyamines, but whose subunits apparently contain two transmembrane domains and a P region. Also distinct from ionotropic neurotransmitter receptors, but sometimes responding to extracellular messages of the same kind, is the "metabotropic" superfamily of monomeric "G-protein-coupled neurotransmitter receptors (NTR)" containing seven transmembrane domains per structural/functional unit. Based on functional criteria, ionotropic neurotransmitter receptors are grouped together as a transmembrane protein superfamily. However, this branch of signaling molecules must in turn have at least three branches based on genetic and structural distinctions. One branch would include P_{2X} receptors for ATP, which have two transmembrane domains and a P region per subunit. Another would include ionotropic glutamate receptors composed of subunits having three transmembrane domains and a P region. The third branch would include ionotropic GABA, glycine, 5-HT_3 and nicotinic receptors, which are likely composed as pentamers of subunits, each of which in turn has four transmembrane domains. Other classifications of some of these signaling molecules have been proposed.[2]

vertebrates. Each of the AChR subunits encoded by these genes is thought to have an extensive N-terminal domain positioned extracellularly, four transmembrane domains (MI-M4) that anchor these integral membrane proteins, and an extracellular C-terminus (see chapter 2 in this volume; see Fig. 7.1).

There is high homology across the family of AChR subunits in the N-terminal extracellular domain and the four transmembrane domains; however, the putative, second intracellular loop between transmembrane domains M3 and M4 is absolutely unique in sequence and sometimes in size for each subunit. Ligand recognition is thought to occur at sites in the N-terminal extracellular domain[3] (see chapter 3 in this volume) and a role for the extracellular loop between M2 and M3 transmembrane domains in coupling of ligand binding to channel gating has been suggested.[4] M2 transmembrane domains from each subunit are thought to form the lining of the ion channel, and the other transmembrane domains are thought to form an outer "crust" for the complex as well as a hydrophobic interface with the lipid membrane. All AChR subunits share with ionotropic glycine, γ-amino butyric acid ($GABA_A$), and serotonin ($5\text{-}HT_3$) receptors expression of cysteine residues 14 amino acids apart in the N-terminal domain (consensus positions 128 and 142 in the amino acid sequences for mature α1-α9).[5] All AChR α subunits also share expression of tandem cysteine residues near the principal agonist binding site on the N-terminal domain (sequence positions 192-193 for α1-α4 and 190-191 for α7-α8).[5]

Amino acid sequences for a given AChR subunit across vertebrates/mammals typically have over 80%/90% identity. Analysis of "informative" amino acid sequences provisionally indicates that AChR subunits fall into four general subfamilies (Fig. 7.2; see chapter 2 in this volume for another perspective).[6] One subfamily includes structural subunits of muscle-type AChR (β1, γ, δ, and ε; note that the latter three subunits also contribute to the ligand-binding pocket). Each of the subunits in a second subfamily harbors sites involved in or influencing ligand binding (α1-α6 and β2-β4). Members in two additional subfamilies of subunits form "neuronal" AChR subtypes that, like muscle-type AChR, are capable of binding α-bungarotoxin (Btx), a curaremimetic neurotoxin from the venom of the Formosan banded krait, *Bungarus multicinctus*. One of these subfamilies includes α7 and α8 subunits, and another includes α9 subunits. Assignment of α1 subunits to the ligand binding subfamily is tentative, because this subunit has a gene structure more like that of the structural subunit subfamily.[6] Placement of β2/β4 subunits is also provisional because it varies somewhat with amino acid sequences sampled (according to Le Novère and Changeux,[6] but not by analyses described in Fig. 7.2), and because informative nucleic acid sequence analyses place β2 and β4 subunits in the structural subunit subfamily (see chapter 2 in this volume).[7] More generally, definitive subunit-subfamily assignments await further technical and conceptual refinement of molecular evolution analyses, potential identification of new AChR subunits, and sequence information for AChR subunits from other species. Most AChR subunit genes have distinct chromosomal localizations, although α3, α5 and β4 subunit genes and γ and δ subunit genes are clustered.[8-9] α3, α5, β4 and γ subunits are more closely related to other subunits than to their chromosomal neighbors, suggesting that formation of the relevant gene clusters by tandem duplications of common ancestors must have occurred prior to translocations that produced α3/α6, α5/β3, β4/β2, and γ/ε pairs.

Subunit Composition of Receptor Subtypes

Diversity of AChR subunits and genes has potentially broad physiological significance. Gene promoter sequences must dictate which AChR subunits are expressed in particular cells, at specific times during development, and respond to signals targeting the nucleus to control AChR expression.[10-13] Probably hidden in the absolutely unique sequences of cytoplasmic domains of each subunit are signals that influence subcellular localization of AChR through interactions with the cytoskeleton[14] and regulate, for example, subunit recognition by kinases and phosphatases that control AChR phosphorylation state.[15] Subtle differences across subunits in amino acids that line the ion channel can have dramatic effects on ion selectivity, conductance, and kinetics of channel opening/closing.[16] Sequences of subunit extracellular domains influence assembly of AChR as oligomers, if extrapolations can be made from studies done on assembly of muscle-type AChR subunits,[17] and effects of extracellular messages on AChR. Sequences of extracellular domains also dictate ligand binding preferences of subunits and AChR subtypes that contain those subunits (see below). Hence, knowledge about their subunit compositions and stoichiometries is essential to our understanding of diverse AChR subtypes.

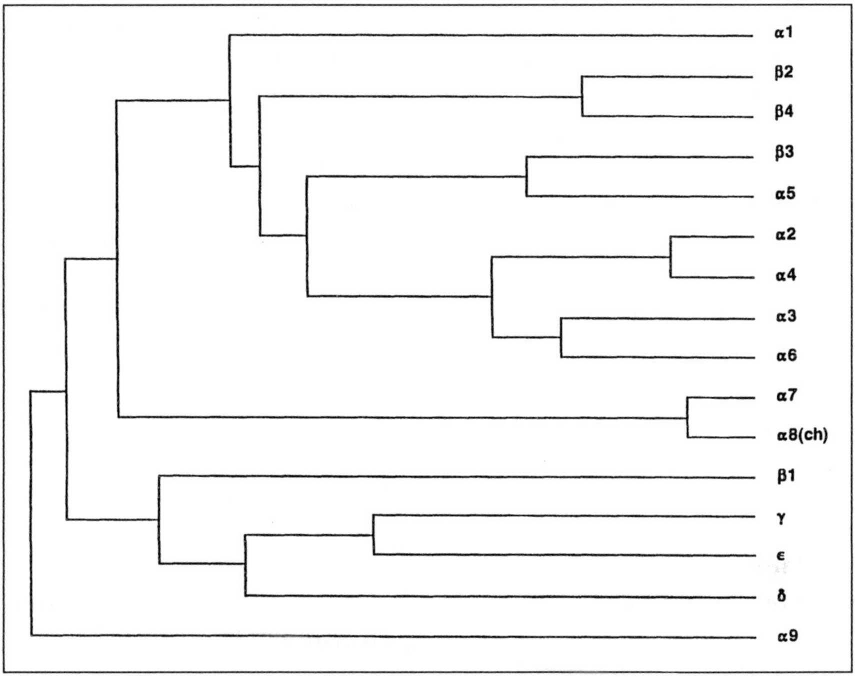

Fig. 7.2. Dendrogram or phylogenetic tree diagram showing relationships between AChR subunits. Full-length amino acid sequences for rat AChR subunits (or for the chick α8 subunit; see Fig. 3 of Lindstrom[5] for sequences) were aligned and subjected to initial analysis using the PCGENE (Intelligenetics) program CLUSTAL set for a k-tuple value of 1, a gap penalty of 5, a window size of 10, a filtering level of 2.5, and open gap and unit gap costs of 10. Sequences representing unique signal peptides and cytoplasmic domains were then deleted, yielding a reference sequence of 408 amino acids (i.e., for the AChR α1 subunit, the sequence used was from SEHETR to STMKRP and from EEWKYV to ELHQQG). Analysis was then repeated using the same parameters and these "informative" amino acid sequences to yield the dendrogram shown. Branch points farthest to the right indicate subunits with closest relationships (about 89% identity plus similarity for α2 and α4 or for α7 and α8), whereas branch points to the left on the diagram indicate lower degrees of identity plus similarity (e.g., about 64% identity plus similarity across β1, γ, δ and ε subunits and from 45-61% identity plus similarity for α9 vs. the other subunits). α5 and β3 sequences fell into a distinct subfamily unless analyses were done omitting cytoplasmic domain sequences, but other relationships were the same. Somewhat different dendrograms were obtained by Ortells and Lunt[7] in analysis of informative nucleic acid sequences (see chapter 2 in this volume) or by Le Novère and Changeux[6] using other analytic approaches (see text).

Some of this information and initial characterization of AChR subtypes has been gained from convergent studies of native AChR in specific tissues or cell lines, of reconstituted AChR in *Xenopus* oocyte expression systems, and of transgenic AChR stably or transiently expressed heterologously in cell lines. Some caution is warranted in evaluating this information, partly because heterologously expressed AChR may not always express properties equivalent to those of native AChR (for both technical and biological reasons; see a recent status report on ionotropic glutamate receptors).[18] The following summary utilizes a nomenclature reliant upon sites of expression in the mature nervous system and known subunit compositions of AChR subtypes. Limited specificity of this nomenclature is acknowledged, given that several subunits and/or functional AChR subtypes have been found to be expressed at different stages of development[19] in tissues as diverse as small cell carcinoma of the lung (and normal pulmonary neuroendocrine cells),[20] keratinocytes,[21] and lymphoid cells,[22] as well as in neurons or muscle, and because available information about AChR subtype subunit composition is clearly incomplete.

For many years, it has been known that vertebrate muscle-type AChR are composed as heteropentamers of two α1 and one each β1, δ and either γ (fetal) or ε (adult) subunits.[23] Oocyte expression studies confirm that this complement of subunits is required for formation of fully functional receptors, although β2 or β4 subunits found in neurons can substitute for β1 subunits.[24] Oocyte expression studies also suggest that pharmacologically and kinetically unique functional channels can be formed more simply as pairwise combinations of α2, α3, or α4 subunits with β2 or β4 subunits,[24-25] sometimes with cross-species differences in pharmacology that are striking given the high conservation of amino acid sequences for AChR subunits.[26] Just as interfaces between α1 and γ/ε or δ subunits are thought to form the ligand binding pocket in muscle-type AChR, interfaces between α2-4 subunits and β2 or β4 subunits appear to constitute ligand binding domains of other AChR subtypes; the pharmacology of formed receptors is strongly influenced by both neuronal α and β subunits. Studies in oocytes also indicate that α6 and β4 subunits can combine to form functional AChR.[27] α7 subunits can combine to form homooligomeric (pentameric) AChR that have many of the properties of native AChR containing α7 subunits when heterologously expressed in transfected cell lines[28-29] or in oocytes.[30] Mutagenesis studies have identified residues on α7 subunits that form not only the "α subunit" complement, but also the "β/γ/ε/δ subunit" complement to ligand binding pockets.[31] Chick α8 subunits also can form functional channels when expressed exogenously alone, but not efficiently, perhaps because appropriate assembly partners are absent.[5] α9 subunits are components of a novel class of AChR that, when heterologously expressed in oocytes as functional homooligomers, recognize muscarine or nicotine (which classically are agonists at muscarinic or nicotinic receptors, respectively) as well as atropine or d-tubocurarine (which classically are antagonists at muscarinic or nicotinic receptors, respectively) as antagonists of ACh-induced channel activity.[32]

From studies of native AChR, one form of vertebrate ganglionic AChR detectable based on functional assays, anti-AChR subunit antibody interactions, and binding of ^{3}H-labeled nicotinic agonists contains α3, α5 and β4 subunits, and a subset of these AChR also contain β2 subunits.[33] Addition of α5 or β2 subunits to AChR containing α3 and β4 subunits heterologously expressed in oocytes alters functional and ligand-binding properties,[34] indicating potential functional relevance of expression of ganglionic AChR more complex than those simply containing α3 and β4 subunits. Another ganglionic AChR subtype detectable based on immunoreactivity, binding of radiolabeled Btx, and functional sensitivity to blockade by Btx contains α7 subunits.[35-36] One vertebrate AChR subtype found in the central nervous system (CNS) also contains α7 subunits,[37] but more work is needed to determine whether native α7-AChR exist as homooligomers and/or as heterooligomers containing additional kinds of subunits.[38] Other CNS Btx-binding AChR forms (expressed in chick but not yet identified in mammals) contain α8 subunits or α8 plus α7 subunits.[39] Chick α7- and chick α8-AChR differ pharmacologically, with chick α8-AChR having generally higher affinities for small nicotinic agonists and lower affinity for the antagonists d-tubocurarine and Btx.[5] The predominant form of mammalian CNS AChR subtype that binds radiolabeled nicotinic agonists such as cytisine, nicotine, epibatidine, or ACh with high affinity contains α4 and β2 subunits.[40-41] Note, however, that there has not yet been a follow-up to the interesting observation that there is an equally plentiful, high-affinity nicotine-binding AChR in chick brain that is distinct from α4β2-AChR.[42] Consistent with findings using muscle-type AChR or heterologously-expressed α7-AChR, α4β2-AChR exist as a pentamer containing two α4 and three β2 subunits.[43-44] Little is known about the composition of native AChR that contain α9 subunits.[32] Definitive assignments of α2, α6 and β3 subunits to specific, naturally-expressed AChR subtypes have not yet been made. Moreover, there may be AChR subunits that have not yet been identified or cloned. Hence, it is likely that there also exist additional AChR subtypes of yet undefined subunit composition. Some of these AChR may already have been partially characterized (e.g., those with features of α3β4-, α3β2-, α3β2β4-, α2β4-AChR, etc.) based on their unique pharmacological profiles when their function in regulation of neurotransmitter release or in excitatory neurotransmission has been studied (see below).

Distributions and Functions of Nicotinic Receptors in Neurons

Each AChR subtype identified to date has a distinctive distribution across and outside of the mature nervous system and unique capacities to recognize nicotinic ligands, and some AChR subtypes might engage in novel functional roles.

Neuronal Nicotinic Systems

Cholinergic Pathways

Mapping of cholinergic pathways based on choline acetyltransferase (ChAT) immunohistochemistry could be considered to define the limits of nicotinic cholinergic signaling systems in the mammalian CNS. History indicates that apparent mismatches between AChR distributions and sites of cholinergic innervation become resolved as techniques improve and searches continue. Thus, these broadly-distributed pathways are candidates for mediation of physiologically-relevant nicotinic cholinergic signaling. ChAT immunohistochemical maps identify major cholinergic projections (and major targets in parentheses) from loosely-delimited nuclei of heterogeneous neurotransmitter phenotype in the medial septum (hippocampus), the nucleus basalis of Meynert (cerebral cortex and amygdala), the ventral nucleus (hippocampus) and horizontal limb (olfactory bulb) of the diagonal band, the pedunculopontine nucleus and the laterodorsal tegmentum of the rostral brainstem (thalamus), the epithalamic medial habenula (interpeduncular nucleus), and the parabigeminal nucleus (superior colliculus).[45] Acetylcholinesterase staining and in situ hybridization using choline acetyltransferase probes largely identify the same areas as major sources (and major targets) for cholinergic innervation.[46] Other projections target structures including the cerebellar cortex and deep cerebellar nuclei, the basal forebrain cholinergic complex, cranial nerve and vestibular nuclei, the reticular formation, the substantia nigra, and the epithalamic medial habenula, which receives dual innervation from both the medial septum/diagonal band and the laterodorsal tegmentum. Intrinsic cholinergic interneurons are present in the caudate-putamen complex and compartments of the ventral striatum, as well as in the retina, although presumably cholinergic local circuit neurons identified (in rodents, but not yet in primates) in the cerebral cortex based on ChAT-like immunoreactivity are not evident based on ChAT mRNA detection.[45-46] ChAT mRNA and immunoreactivity are also present in motor and parasympathetic nuclei associated with cranial nerves 3-12 (including cochlear and vestibular efferent nuclei), in α- and γ-motoneurons throughout the ventral horn of the spinal cord, and in thoracic and lumbar intermediolateral cell columns.[46] Ultrastructural analyses suggest that at least some of the projections end in classical synaptic contacts, but do not exclude the possibility of extrasynaptic release.[45]

AChR Distribution

Mapping of AChR distributions at low resolution based on radioligand binding autoradiography is consistent with expression of some form of AChR in most of these "major" or "minor" cholinergic targets[47-48] as well as in some sources of major cholinergic projections that also happen to be cholinergic targets (e.g., nucleus basalis and medial habenula). Anatomic analyses also suggest that presumably $\alpha 4\beta 2$-AChR and $\alpha 7$-AChR (labeled using ^{3}H-labeled nicotinic agonists and ^{125}I-labeled Btx, respectively) have largely unique but sometimes overlapping distributions.[47-48] For example, in the cerebral cortex, ^{3}H-labeled agonist binding sites are most prominent in layers III/IV, but occur in all layers, whereas ^{125}I-labeled Btx binding sites are most concentrated in layers V and VI. The medial habenula and the thalamus contain radiolabeled agonist binding sites but not sites for Btx, whereas the interpeduncular nucleus and the medial septum possess high densities of both types of sites, and Btx sites predominate in the hypothalamus and hippocampus.

Immunocytochemical studies, which generally offer higher resolution and sensitivity than ligand binding autoradiographic studies, suggest that β2 subunits are expressed quite widely in the brain, even at sites where α4 subunits and/or high affinity agonist binding-AChR have not been found. Subtle differences in distributions of β2-like antigens are found even across immunocytochemical studies, perhaps reflecting use of fresh or fixed tissue or differing specificities of antibodies used (e.g., compare Britto et al,[49] Hill et al,[50] to Swanson et al[51]), but perhaps also realistically reflecting possibly important differences in accessibility/fragility of β2 epitopes on axons and their termini ("presynaptic," "preterminal") as opposed to on dendrites and cell bodies ("postsynaptic"). Anti-α4 subunit immunoreactivity should closely correlate with that of high-affinity radioagonist binding sites, but published accounts indicate presence of α4 subunits in brain regions (hypothalamus, hippocampus) without a high density of high affinity agonist binding sites.[52] As might be expected, anti-α7 subunit immunoreactivity has a distribution encompassing and sometimes exceeding (some thalamic nuclei, cerebellum) that predicted from autoradiographic studies employing radiolabeled Btx.[53] However, much more work is needed to determine the location of these sites at the ultrastructural level, particularly given technical/analytical limitations in some of the earlier electron-microscopic/autoradiographic studies. Available anatomic and lesioning studies are consistent with a "presynaptic" disposition of AChR sites that bind nicotine with high affinity or react with anti-β2 subunit antibodies.[54] On the other hand, anti-α4 subunit immunostaining in the hypothalamus is consistent with both pre- and postsynaptic dispositions,[52] whereas most studies indicate that α7-AChR are located "postsynaptically."[53-54]

In situ hybridization-based maps at high resolution and sensitivity of AChR subunit messages do not always overlap with maps of subunit antigens or radioligand binding sites, probably due to transport of subunits/AChR from sites of their synthesis. Nevertheless, these studies generally indicate very broad distribution of β2 subunit mRNA, and narrower but still wide distribution of α4 or α7 subunit mRNAs, at sites largely consistent with observations made in immunocytochemical or autoradiographic studies.[47-48,54] There are much more restricted distributions to only a few nuclei of α2 (interpeduncular nucleus, weak in deep cortical layers), α5 (interpeduncular nucleus, substantia nigra, ventral tegmentum, cortical layer

VIb), and β3 (medial habenula, substantia nigra, ventral tegmentum), subunit mRNAs.[47-48,55] By contrast, α3 (medial habenula, substantia nigra, ventral tegmentum, locus coeruleus, thalamus, strong in cortical layer IV) and β4 subunit messages (medial habenula; cortical layer IV, interpeduncular nucleus, hippocampal formation) have intermediate ranges of distribution.[47-48,55-56] Cells in some brain regions not identified as having strong cholinergic innervation express AChR transcripts (e.g., Purkinje cells of the cerebellum).[25,57]

The retina is very rich in a large variety of AChR subtypes and subunits,[39] and Btx-binding sites as well as α7 and α9 subunits are found in the cochlea.[32,58]

Radioligand binding, immunocytochemical, and/or in situ hybridization studies clearly show expression of muscle-type AChR and their subunits postsynaptically at the nerve-muscle junction, of ganglionic α3β4α5-AChR on postganglionic neuronal somatodendritic fields, and of ganglionic α7-AChR mostly at perisynaptic sites.[59] AChR might also be expected to exist on motoneuronal terminals, preganglionic terminals, and selected autonomic (parasympathetic, some sympathetic) terminals/targets.

Functional Roles

As generalizations drawn from studies of heterologously-expressed or native AChR,[24,37,60] AChR in neurons have comparatively high permeabilities to Ca^{2+} and channels subject to a unique form of Mg^{2+}-mediated block responsible for rectification of current at positive membrane potentials. Many AChR subtypes in CNS neurons also exhibit fast kinetics for functional inactivation that compromises attempts to study these AChR using electrophysiological techniques, perhaps contributing to previous impressions that nicotinic cholinergic signaling is not nearly as prominent as muscarinic cholinergic signaling in the CNS.[61] Specific electrical stimulation of cholinergic fibers is also difficult to achieve in the CNS; consequently, much of the understanding of nicotinic cholinergic signaling relies on "electropharmacological" approaches involving electrical measures of neuronal responses to exogenously applied ACh or other nicotinic agents.

Classical Excitatory Neurotransmission

Muscle-type AChR containing α1, β1, δ and γ subunits embryonically or ε subunits in the adult, and ganglionic AChR now known to contain α3, β4 and α5 subunits, represent both classical and contemporary models for the establishment of concepts pertaining to mechanisms of drug action, synaptic transmission, and structure, function, and diversity of transmembrane signaling molecules. These are the best characterized AChR, and they mediate depolarizing, inward Na^{+} currents involved in classical excitatory neurotransmission at the nerve-muscle junction and through autonomic ganglia, respectively. Despite a predominant perisynaptic localization, and their prominent Ca^{2+} permeability, α7-AChR also contribute substantially to ganglionic synaptic currents.[62]

There also is excellent evidence for actions of AChR in the mediation of excitatory neurotransmission at some sites in the CNS where they might contribute to nicotine-sensitive processes involved in emotion, sensory processing, and cognition. Nicotinic excitatory neurotransmission occurs not only at motoneuronal-Renshaw cell synapses, but also in the thalamus[63] and nucleus ambiguus.[64] AChR in the thalamus possibly involved in excitatory neurotransmission can be detected using ion flux assays and synaptosomal preparations.[65] AChR functional channel responses to applied agonists are detectable in hippocampal neurons (reflecting expression of a heterogenous collection of AChR deduced from comparisons to oocyte expression studies to contain α4 and β2, α3 and β4, or α7 subunits) and in selected other central sites.[37,47,54,60]

Local Control of Neurotransmitter Release

Technical improvements allowing the detection of transient and novel functional responses of AChR as well as advances in the ability to identify ions that permeate AChR channels have led to the recognition that AChR can play roles other than mediation of classical excitatory neurotransmission. For example, AChR composed as homooligomers of α7 subunits expressed in *Xenopus* oocytes or native AChR containing α7 subunits bind curaremimetic neurotoxins and mediate very short-lived, nicotine-gated, ion channel responses of high Ca^{2+} permeability rivaling or surpassing that of the NMDA class of ionotropic glutamate receptors (iGluR).[5,37] Other CNS AChR also have significant Ca^{2+} permeability.[60,66] Consistent not only with their permeability to Ca^{2+}, but also with their "preterminal" location on axons or their "presynaptic" location on axon terminals, electrophysiological studies and investigations of release of neurotransmitters from brain synaptosomes or tissue slices, or even in the periphery from motoneuronal terminals,[67] indicate that some AChR subtypes appear to be involved in control of neurotransmitter release.[60,67-73] Locations of "presynaptic" AChR (and the types of neurotransmitters whose release they modulate) include the hippocampus (acetylcholine, norepinephrine, GABA, 5-HT), the striatum and nucleus accumbens (dopamine), the interpeduncular nucleus (GABA, glutamate), the lateral geniculate nucleus (GABA), the cerebral cortex (acetylcholine, glutamate, GABA), and the cerebellum (GABA; op. cit.). For many years, AChR biologists were concerned with their inability to detect excitatory neurotransmission mediated by AChR at pathways thought to be nicotinic cholinergic based on staining for choline acetyltransferase, acetylcholinesterase, and AChR. Now, it seems possible that release of ACh at those sites modulates release of other neurotransmitters, perhaps without requiring action potential propagation from the corresponding cell bodies. Hence, a good proportion of nicotinic cholinergic signaling is not simply involved in completion of neuronal circuits, but rather in the regulation of the neurochemical "soup"[73] that bathes the brain and influences neuronal connectivity. This type of regulatory function would be consistent with newly discovered roles of AChR in processes such as long-term potentiation[75-76] and with the comparatively innocuous effects of chronic nicotine exposure or AChR gene mutations or knockouts on nervous system function (see below).[77-78] If ambient levels of ACh (or nicotine in tobacco users) in the brain are high enough to affect balance between activation and desensitization of AChR (see below), then variations in ACh (locally or at a distance from release sites) or nicotine concentration could fine-tune activity of AChR and neuronal circuitry or signaling.

Modification of Neuronal Architecture

Another novel role of AChR consistent with their Ca^{2+} permeability, their discovery at some sites before synapse maturation occurs, and their presence on dendrites and axon terminals, is in structuring and maintenance of neurites and synapses, i.e., in pathfinding and target detection. There is both long-standing[79] and more recent[80-82] evidence that nicotine, nicotinic agonists, or nicotinic antagonists such as Btx can influence neurite outgrowth in vitro and synaptogenesis in vivo. Thus, regulation of axonal growth cone or dendrite extension/retraction represents yet another way that ACh, release of which might fluctuate as growth cones/axons and dendrites make contact, or exogenous nicotine can influence neuronal circuitry and nervous system activity. Reciprocally, studies linking integrity of the cytoskeleton with levels of AChR expression provide a potential mechanism for feedback control of AChR-mediated changes in neurite structure.[83]

Roles in Neuronal Death and Survival

Just as Ca^{2+}-permeable iGluR have been implicated in neurotoxicity, activity of AChR could also affect neuronal viability/death.[84] In particular, recent studies strongly implicate spinal AChR in mediation of the effects of nicotinic ligands on motoneuronal death.[85-86] A form of spontaneously-occurring, late-onset death of a subset of neurons in the nematode[87] is caused by a mutation in a AChR analogous to a mutation that causes a loss in desensitization of Ca^{2+}-permeable, vertebrate α7-AChR.[16] Further work is needed to determine whether any forms of spontaneous neurodegeneration occur in vertebrates expressing similarly mutated AChR subunits. Chronic nicotine treatment is also "neuroprotective" in several models of neurodegenerative diseases.[88-90] Neuronal survival could be influenced by perturbation of AChR involved in excitatory neurotransmission, release of cytokines or growth factors, activation of systems such as mitochondrial oxidative respiration, modulation of neurotransmitter release, and/or structuring of neuronal connections.

Mediation of Signaling via Other Neuroactive Agents

Aside from altering transmembrane potentials and Ca^{2+} permeability, AChR also may exert their effects by altering metabolism of agents such as nitric oxide[91] and activating neuroendocrine systems and/or gene expression.[92] Recent studies also suggest that AChR, like iGluR and $GABA_A$-R, may be targets for natural, pharmacological, and/or recreational modulation of nervous system function mediated by a variety of agents including steroids,[93] peptides,[94] anesthetic agents,[95] hallucinogenic agents and imidazolines,[96] and calcium channel blockers.[97]

Nicotinic Receptors and Molecular Bases for Nicotine Dependence

Nicotine has been suggested to represent a model dependent substance, and AChR must be involved in mediation of its effects on nervous system function.

Behavioral Effects, Tolerance and Dependence

Nicotine exposure at different doses and for different periods of time has a number of physiological and behavioral effects in laboratory animals and humans.[98-101] These effects range from induction of locomotor activity, seizures and changes in body temperature to relief of anxiety, depression, or pain and enhancement of attention and cognition. As the biologically active substance implicated in "addiction" to tobacco products, nicotine might also elicit activation of pleasure/reward centers in the brain, induce effects that account for nicotine dependence and tolerance, and contribute to the unpleasant effects of nicotine withdrawal. Chronic use of nicotine shares with other addictive processes manifestation of craving, tolerance, physical and psychological (mild euphoriant) dependence, relapse during abstinence, and withdrawal symptoms (op. cit.). Nicotine is not popularly viewed, as are narcotics, as an intoxicating and/or performance/judgment-altering drug of abuse, consumption of which endangers the user and/or other members of society. However, nicotine-dependent tobacco consumption contributes to health problems in a population much larger than the population of narcotic users and at higher costs.[102]

Effects of Chronic Nicotine Exposure on Nicotinic Receptor Numbers and Function

A description of how nicotine exposure affects nervous system function requires an improved understanding of effects of nicotine exposure on numbers and function of AChR, which ultimately mediate actions of nicotine. This information is critical to an understanding of cellular and molecular bases for nicotine dependence and the design of effective strategies to treat or eliminate use of tobacco products and exposure to its noxious constituents, even if such strategies involve pharmacological intervention to mimic the perceived beneficial or pleasurable effects of nicotine without engaging mechanisms leading to addiction. Acute exposure of naive subjects or experimental systems to nicotine is expected to activate AChR and nicotinic cholinergic signaling and could account for some effects of nicotine action. However, longer-lasting exposure to nicotine has different effects. Among these is a rapid in onset and reversible phase of AChR functional loss called "desensitization." An effect of more chronic nicotine exposure that has gained considerable attention is the numerical upregulation of AChR-like radioligand binding and antigenic sites in the CNS, both in intact animals, including man, and in many types of experimental system.[41,48,103-106] This response exhibits some heterogeneity across AChR subtypes, even after accounting for interlaboratory technical variations in study design and execution.[106-107] It has been postulated that AChR upregulation is compensatory for the loss in AChR function due to desensitization, but such an adaptive response does not involve nuclear mechanisms and changes in steady state levels of AChR subunit mRNA, although AChR subtype-specific effects on stability and/or degradation of AChR or their precursors appears to be involved.[48,103,105-106,108] Chronic nicotine exposure also induces in many experimental systems a persistent inactivation of AChR function that is distinct from desensitization, exhibits heterogeneity across AChR subtypes,[103,106,109-110] and occurs for α4β2-AChR at concentrations of nicotine found at steady state in the plasma of smokers (~200 nM). Evidence from a variety of approaches suggests that upregulation and persistent inactivation could be mechanistically and causally distinct.[106] It remains to be seen whether the physiologically-relevant effect of habitual use of tobacco products involves contributions from acute activation and upregulation of AChR or rather is dominated by persistent inactivation-induced disabling of nicotinic cholinergic signaling,[106-107] perhaps in circuits that when hyperactive contribute to anxiety and compromised attention and cognition. The latter view is consistent with experiments using transgenic mice lacking expression of β2 subunits and high-affinity nicotine binding sites and whose performance in a passive avoidance test was enhanced relative to that of wild-type mice but comparable to that of wild-type mice who had been treated with nicotine.[78] There also is evidence that behavioral tolerance to nicotine is related to inactivation of AChR.[111] However, as more AChR subunits and subtypes are identified, the possibilities multiply. For example, nicotine acts acutely as an antagonist at AChR containing α9[32] and as a partial agonist or antagonist at AChR thought to be localized to "reward" centers of the brain and containing α6 plus β4 or α3 plus β2 subunits.[27] Thus, nicotine's effects are likely to reflect its temporally-integrated, AChR subtype- and brain region-specific acute actions as an agonist, partial agonist, or antagonist in the more chronic induction of desensitization and persistent inactivation. Moreover, it is possible that long-term changes in nervous system function after chronic exposure to nicotine involve changes in neuronal architecture, survival, and metabolism, all of which might contribute to nicotine tolerance, dependence, and symptoms of withdrawal during abstinence.[106]

Prospects

Particularly when it is realized that the concept of AChR diversity was reborn as genetically-based only over the last decade or so, and given the accelerating rate of technical innovation applicable to studies of AChR, one should not be surprised if current views of AChR and nicotinic signaling will be judged to be incredibly naive ten years hence. Nevertheless, substantial advances in our understanding of AChR and their biological roles are anticipated in several fronts.

Structure of AChR and Their Roles in Nervous System Function

Mutagenesis studies using heterologous expression systems are expected to provide an increasingly clearer picture of the amino acid residues that constitute ligand binding, subunit assembly, and channel interfaces of AChR,[3,16,24,31,112] thereby providing important new insights regarding basic principles of neurotransmitter-gated ion channel structure and function. Recent technical and strategic innovations in electrophysiologically-based approaches applicable to studies of native AChR on neuronal dendrites, soma, or terminals[113] and heterologously expressed AChR of defined subunit composition in oocytes[66] offer promise that functional AChR will be characterized more expeditiously than in the past, potentially providing new views of how nicotinic signaling affects nervous system function. Also expected is increased attention on roles played by AChR not just in synaptic function, but also, possibly via novel signaling cascades, in control of gene expression, in modulation of cytokine activity and release, and in regulation of subcellular functions such as mitochondrial respiration.

Receptor Subtype-Specific Nicotinic Drug Design

Several recent reviews/reports have presented perspectives on how AChR-targeted compounds might be developed that could have some of the positive effects of nicotine (anxiolysis, anti-depression, cognitive enhancement) without requiring delivery of the agent via tobacco products and without the attendant negative health consequences of tobacco usage.[70,114-116] Particularly if there are AChR subtypes that are more significantly involved in the development of nicotine tolerance and/or withdrawal, dependence on nicotine might be avoided if analogs can be devised that have selectivity for other AChR subtypes. Delivery of nicotine via chewing gum or transdermal patches has been used to help in the cessation of tobacco use, and therapeutic nicotine or its analogs could be delivered similarly, although physiologically-relevant drug tissue distributions and concentrations ("nicotine boost") achieved using these systems are not precisely like those realized during tobacco product use.[117] However, alternative delivery systems such as those that would produce vapors rich in nicotine (nasal sprays or nicotine inhalers) offer novel means for nicotine or analog delivery that have attractive pharmacokinetic properties (op. cit.). Also driving the potential therapeutic use of nicotine and its analogs is a variety of studies indicating (sometimes dramatic) beneficial effects of nicotine in the relief of symptoms in Tourette's syndrome,[118] in Parkinson's or Alzheimer's disease patients,[119] in schizophrenics,[120] in adults with attention deficit/hyperactivity disorder,[121] as well as in peripheral disorders such as ulcerative colitis[122] and perhaps acantholysis.[21] Intriguing are consistent findings of decreased AChR in Parkinson's disease[123] and of a negative correlation between smoking and incidence of Parkinson's disease.[124] More controversial are suggestions of a negative correlation between smoking and the incidence of Alzheimer's disease,[125] but evidence for decreased AChR in Alzheimer's disease has accumulated.[126] Efforts to test for causal relationships between nicotine exposure, AChR function, and neurodegenerative diseases can be anticipated. If nothing else, clinically-driven studies of nicotine, its analogs, and agents such as epibatidine[127-128] will ensure development of novel and useful tools for characterization and classification of diverse AChR subtypes.

Is There Individual Variation in Receptor Expression?

Important findings relevant to the biological roles of AChR concern identification of a variety of diseases that are associated with mutations in AChR subunits and/or, for example, autoimmune responses against diverse AChR subtypes (myasthenia gravis;[129-131] forms of epilepsy;[77] acantholysis;[21] see chapter 8 in this volume on other pathological conditions affecting AChR). In the next decade, AChR medical biology might be expected to reveal that individuals of differing AChR genotypes have differing predilections to initiation, continuation, or ability to cease use of tobacco products.[132] Perhaps behavioral attributes of individuals, such as high or blunted anxiety or even cognitive or learning disabilities, might be influenced by AChR genotype and diagnosable based on acute sensitivity to nicotinic agents.

Acknowledgments

Work in the author's laboratory has been supported, at different stages, by grants from the National Institute on Drug Abuse (DA07319), the National Institute of Neurological Disorders and Stroke (NS16821), the Smokeless Tobacco Research Council (0277-01), the Arizona Disease Control Research Commission (82-1-098, 9615, and 9730), the American Parkinson Disease Association, and the Council for Tobacco Research—U.S.A. (1683A and 4366M), by Epi-Hab Phoenix, Inc., and by faculty endowment and laboratory capitalization funds from the Men's and Women's Boards of the Barrow Neurological Foundation. The contents of this report are solely the responsibility of the author and do not necessarily represent the views of the aforementioned awarding agencies.

References

1. Lukas RJ, Bencherif M. Heterogeneity and regulation of nicotinic acetylcholine receptors. Intl Rev Neurobiol 1992; 34:25-131.
2. Barnard EA. The transmitter-gated channels: A range of receptor types and structures. Trends Pharm Sci 1996; 17:305-309.
3. Galzi J-L, Changeux J-P. Ligand-gated ion channels as unconventional allosteric proteins. Curr Opin Struct Biol 1994; 4:554-565.
4. Campos-Caro A, Sala S, Ballesta JJ, Vicente-Agullo F, Criado M, Sala F. A single residue in the M2-M3 loop is a major determinant of coupling between binding and gating in neuronal nicotinic receptors. Proc Natl Acad Sci USA 1996; 93:6118-6123.

5. Lindstrom J. Neuronal nicotinic acetylcholine receptors. In: T Narahashi, ed. Ion Channels, Vol. 4. New York: Plenum Press, 1996:377-450.
6. Le Novère N, Changeux J-P. Molecular evolution of the nicotinic acetylcholine receptor: An example of multigene family in excitable cells. J Molec Evol 1995; 40:155-172.
7. Ortells MO, Lunt GG. Evolutionary history of the ligand-gated ion channel superfamily of receptors. Trends Neurol Sci 1995; 18:121-127.
8. Anand R, Lindstrom J. Chromosomal localization of seven neuronal nicotinic receptor subunit genes in humans. Genomics 1992; 13:962-967.
9. Orr-Urtreger A, Seldin MF, Baldini A, Beaudet AL. Cloning and mapping of the mouse α7-neuronal nicotinic acetylcholine receptor. Genomics 1995; 26:399-402.
10. Bessis A, Savatier N, Devillers-Thiery A, Benjamin A, Changeux J-P. Negative regulatory elements upstream of a novel exon of the neuronal nicotinic acetylcholine receptor α2 subunit gene. Nucleic Acids Res 1993; 21:2185-2192.
11. Daubas P, Salmon AM, Zoli M, Geoffroy B, Devillers-Thiéry A, Bessis A, Médevielle F, Changeux J-P. Chicken neuronal acetylcholine receptor α2-subunit gene exhibits neuron-specific expression in the brain and spinal cord of transgenic mice. Proc Natl Acad Sci USA 1993; 90:2237-2241.
12. Hamassaki-Britto D, Gardino PF, Hokoc JN, Keyser KT, Karten HJ, Lindstrom JM, Britto LR. Differential development of α-bungarotoxin-sensitive and α-bungarotoxin-insensitive nicotinic acetylcholine receptors in the chick retina. J Comp Neurol 1994; 347:161-170.
13. Hernandez MC, Erkman L, Matter-Sadzinski L, Roztocil T, Ballivet M, Matter JM. Characterization of the nicotinic acetylcholine receptor β3 gene. J Biol Chem 1995; 270:3224-3233.
14. Froehner S. Regulation of ion channel distribution at synapses. Annu Rev Neurosci 1993; 16:347-368.
15. Nakayama H, Okuda H, Nakashima T. Phosphorylation of rat brain nicotinic acetylcholine receptor by cAMP-dependent protein kinase in vitro. Molec Brain Res 1993; 20:171-177.
16. Bertrand D, Galzi J-L, Devillers-Thiery A, Bertrand S, Changeux J-P. Stratification of the channel domain in neurotransmitter receptors. Curr Opin Cell Biol 1993; 5:688-693.
17. Kreinkamp HJ, Maeda R, Sine S, Taylor P. Intersubunit contacts governing assembly of the mammalian nicotinic acetylcholine receptor. Neuron 1995; 14:635-644.
18. Sucher NJ, Awobuluyi M, Choi Y-B, Lipton SA. NMDA receptors: from genes to channels. Trends Pharm Sci 1996; 17:348-355.
19. Corriveau R, Romano S, Conroy W, Olivia L, Berg D. Expression of neuronal acetylcholine receptor genes in vertebrate skeletal muscle during development. J Neurosci 1995; 15:1372-1383.

20. Schuller HM. Mechanisms of nicotine stimulated cell proliferation in normal and neoplastic neuroendocrine lung cells. In: Clarke PBS, Quik M, Adlkofer F, Thurau K, eds. Effects of Nicotine on Biological Systems II. Basel: Birkhäuser Verlag, 1995:151-158.
21. Grando SA, Horton RM, Pereira EF, Diethelm-Okita BM, George PM, Albuquerque EX, Conti-Fine BM. A nicotinic acetylcholine receptor regulating cell adhesion and motility is expressed in human keratinocytes. J Invest Dermatol 1995; 105:774-781.
22. Mihovilovic M, Roses AD. Expression of alpha-3, alpha-5, and beta-4 neuronal acetylcholine receptor subunit transcripts in normal and myasthenia gravis thymus: Identification of thymocytes expressing the alpha-3 transcripts. J Immunol 1993; 151:6517-6524.
23. Karlin A, Akabas MH. Toward a structural basis for the function of nicotinic acetylcholine receptors and their cousins. Neuron 1995; 15:1231-1244.
24. Papke R. The kinetic properties of neuronal nicotinic receptor. Genetic basis of functional diversity. Prog Neurobiol 1993; 41:509-531.
25. Patrick J, Séguéla P, Vernino S, Amador M, Luetje C, Dani JA. Functional diversity of neuronal nicotinic acetylcholine receptors. Prog Brain Res 1993; 98:113-120.
26. Gerzanich V, Peng X, Wang F, Wells G, Anand R, Fletcher S, Lindstrom J. Comparative pharmacology of epibatidine, a potent agonist for neuronal nicotinic acetylcholine receptors. Mol Pharm 1995; 48:774-782.
27. Gerzanich V, Kuryatov A, Anand R, Lindstrom J. "Orphan" α6 nicotinic AChR subunit can form a functional heteromeric acetylcholine receptor. Mol Pharm 1997; 51:320-327.
28. Puchacz E, Buisson B, Bertrand D, Lukas RJ. Functional expression of nicotinic acetylcholine receptors containing rat α7 subunits in human neuroblastoma cells. FEBS Letters 1994; 354:155-159.
29. Quik M, Choremis J, Komourian J, Lukas RJ, Puchacz E. Similarity between rat brain nicotinic α-bungarotoxin receptors and stably expressed α-bungarotoxin binding sites. J Neurochem 1996; 67:145-154.
30. Palma E, Bertrand S, Binzoni T, Bertrand D. Neuronal nicotinic α7 receptor expressed in *Xenopus* oocytes presents five putative binding sites for methyllycaconitine. J Physiol 1996; 491.1:151-161.
31. Corringer PJ, Galzi J-L, Eisele J-L, Bertrand S, Changeux J-P, Bertrand D. Identification of a new component of the agonist binding site of the nicotinic α7 homooligomeric receptor. J Biol Chem 1995; 270:11749-11752.
32. Elgoyhen AB, Johnson DS, Boulter J, Vetter DE, Heinemann S. α9: An acetylcholine receptor with novel pharmacological properties expressed in rat cochlear hair cells. Cell 1994; 79:705-715.
33. Conroy W, Berg D. Neurons can maintain multiple classes of nicotinic acetylcholine receptors distinguished by different subunit compositions. J Biol Chem 1995; 270:4424-4431.

34. Wang F, Gerzanich V, Wells GB, Anand R, Peng X, Keyser K, Lindstrom J. Assembly of human neuronal nicotinic receptor α5 subunits with α3, β2, and β4 subunits. J Biol Chem 1996; 271: 17656-17665.
35. Lukas RJ, Norman SA, Lucero L. Characterization of nicotinic acetylcholine receptors expressed by cells of the SH-SY5Y human neuroblastoma clonal line. Molec Cell Neurosci 1993; 4:1-12.
36. Vernallis A, Conroy W, Berg D. Neurons assemble acetylcholine receptors with as many as three kinds of subunits while maintaining subunit segregation among receptor subtypes. Neuron 1993; 10: 451-464.
37. Albuquerque EX, Pereira EFR, Castro NG, Alkondon M, Reinhardt S, Schröder H, Maelicke A. Nicotinic receptor function in the mammalian central nervous system. Ann NY Acad Sci 1995; 757:48-72.
38. McGehee DS, Heath MJS, Gelber S, Devay P, Role LW. Nicotine enhancement of fast excitatory synaptic transmission in CNS by presynaptic receptors. Science 1995; 269:1692-1696.
39. Keyser KT, Britto LRG, Schoepfer R, Whiting P, Cooper J, Conroy W, Brozozowska-Prechtl A, Karten JH, Lindstrom J. Three subtypes of α-bungarotoxin-sensitive nicotinic acetylcholine receptors are expressed in chick retina. J Neurosci 1993; 13:442-454.
40. Whiting PJ, Lindstrom J. Purification and characterization of a nicotinic acetylcholine receptor from rat brain. Proc Natl Acad Sci USA 1987; 84:595-599.
41. Flores CM, Rogers SW, Pabreza LA, Wolfe BB, Kellar KJ. A subtype of nicotinic cholinergic receptor in rat brain is composed of α4 and β2 subunits and is up-regulated by chronic nicotine treatment. Mol Pharm 1992; 41:31-37.
42. Whiting P, Liu R, Morley B, Lindstrom J. Structurally different neuronal nicotinic acetylcholine receptor subtypes purified and characterized using monoclonal antibodies. J Neurosci 1987; 7:4005-4016.
43. Anand R, Conroy WG, Schoepfer R, Whiting P, Lindstrom J. Chicken neuronal nicotinic acetylcholine receptors expressed in Xenopus oocytes have a pentameric quaternary structure. J Biol Chem 1991; 266:11192-11198.
44. Cooper E, Couturier S, Ballivet M. Pentameric structure and subunit stoichiometry of a neuronal nicotinic acetylcholine receptor. Nature 1991; 350:235-238.
45. Mesulam M-M. Cholinergic pathways and the ascending reticular activating system of the human brain. Ann NY Acad Sci 1995; 757:169-179.
46. Butcher LL, Oh JD, Woolf NJ. Cholinergic neurons identified by in situ hybridization biochemistry. Prog Brain Res 1993; 98:1-8.
47. Clarke PBS. Nicotinic receptors in mammalian brain: localization and relation to cholinergic function. Prog Brain Res 1993; 98:77-83.

48. Marks MJ, Pauly JR, Gross SD, Deneris ES, Hermans-Borgemeyer I, Heinemann S, Collins AC. Nicotine binding and nicotinic receptor subunit RNA after chronic nicotine treatment. J Neurosci 1992; 12:2765-2784.
49. Britto LRG, Keyser KT, Lindstrom JM, Karten HJ. Immunohistochemical localization of nicotinic acetylcholine receptor subunits in the mesencephalon and diencephalon of the chick (*Gallus gallus*). J Comp Neurol 1992; 317:325-340.
50. Hill JA Jr, Zoli M, Bourgeois J-P, Changeux J-P. Immunocytochemical localization of a neuronal nicotinic receptor. The β2-subunit. J Neurosci 1993; 13:1551-1568.
51. Swanson LW, Simmons DM, Whiting PJ, Lindstrom J. Immunohistochemical localization of neuronal nicotinic receptors in the rodent central nervous system. J Neurosci 1987; 7:3334-3342.
52. Okuda H, Shioda S, Nakai Y, Nakayama H, Okamoto M, Nakashima T. Immunocytochemical localization of nicotinic acetylcholine receptor in rat hypothalamus. Brain Res 1993; 625:145-151.
53. Dominguez del Toro E, Juiz JM, Peng X, Lindstrom J, Criado M. Immunocytochemical localization of the α7 subunit of the nicotinic acetylcholine receptor in the rat central nervous system. J Comp Neurol 1994; 349:325-342.
54. Clarke PBS. Nicotinic receptors and cholinergic neurotransmission in the central nervous system. Ann NY Acad Sci 1995; 757:73-83.
55. Wada E, McKinnon D, Heinemann S, Patrick J, Swanson LW. The distribution of mRNA encoded by a new member of the neuronal nicotinic acetylcholine receptor gene family (α5) in the rat central nervous system. Brain Res 1990; 526:45-53.
56. Dineley-Miller K, Patrick J. Gene transcripts for the nicotinic acetylcholine receptor subunit, Beta4, are distributed in multiple areas of the rat central nervous system. Molec Brain Res 1992; 16:339-344.
57. Zoli M, Le Novère N, Hill JA, Changeux J-P. Developmental regulation of nicotinic receptor subunit mRNAs in the rat central and peripheral nervous system. J Neurosci 1995; 15:1912-1939.
58. Fuchs PA, Murrow BW. A novel cholinergic receptor mediates inhibition of chick cochlear hair cells. Proc Roy Soc Lond 1992; B248:35-40.
59. Wilson Horch HL, Sargent PB. Perisynaptic surface distribution of multiple classes of nicotinic acetylcholine receptors on neurons in the chick ciliary ganglion. J Neurosci 1995; 15:7778-7795.
60. Mulle C, Léna C, Changeux J-P. Electrophysiology of neuronal nicotinic receptors in the CNS. In: Clarke PBS, Quik M, Adlkofer F, Thurau K, eds. Effects of Nicotine on Biological Systems II. Basel: Birkhäuser Verlag, 1995:127-135.
61. Krnjevi´c K. Central cholinergic mechanisms and function. Prog Brain Res 1993; 98:285-292.

62. Zhang Z-w, Coggan JS, Berg DK. Synaptic currents generated by neuronal acetylcholine receptors sensitive to α-bungarotoxin. Neuron 1996; 17:1231-1240.
63. Curro Dossi R, Paré D, Steriade M. Short-lasting nicotinic and long-lasting muscarinic depolarizing responses of thalamocortical neurons to stimulation of mesopontine cholinergic nuclei. J Neurophysiol 1991; 65:393-405.
64. Zhang M, Wang YT, Vyas DM, Neuman RS, Bieger D. Nicotinic cholinoceptor-mediated excitatory postsynaptic potentials in rat nucleus ambiguus. Exp Brain Res 1993; 96:83-88.
65. Marks MJ, Robinson SF, Collins AC. Nicotinic agonists differ in activation and desensitization of $^{86}Rb^+$ efflux from mouse thalamic synaptosomes. J Pharm Exper 1996; 277:1383-1396.
66. Costa A, Patrick J, Dani J. Improved technique for studying ion channel expression in *Xenopus* oocytes, including fast perfusion. Biophys J 1994; 67:1-7.
67. Bowman WC, Marshall IG, Gibb AJ, Harborne AJ. Feedback control of transmitter release at the neuromuscular junction. Trends Pharm Sci 1988; 9:16-20.
68. McGehee D, Role L. Physiological diversity of nicotinic acetylcholine receptors expressed by vertebrate neurons. Annu Rev Physiol 1995; 57:521-546.
69. Grady S, Marks MJ, Collins AC. Desensitization of nicotine-stimulated ^3H-dopamine release from mouse striatal synaptosomes. J Neurochem 1994; 62:1390-1398.
70. Sacaan AI, Dunlop JL, Lloyd GK. Pharmacological characterization of neuronal acetylcholine gated ion channel receptor-mediated hippocampal norepinephrine and striatal dopamine release from rat brain slices. J Pharm Exper Thera 1995; 274:224-230.
71. Sershen H, Toth A, Lajtha A, Vizi ES. Nicotine effects on presynaptic receptor interactions. Ann NY Acad Sci 1995; 757:238-244.
72. Vizi ES, Sershen H, Balla A, Mike Á, Windisch K, Jurányi ZS, Lajtha A. Neurochemical evidence of heterogeneity of presynaptic and somatodendritic nicotinic acetylcholine receptors. Ann NY Acad Sci 1995; 757:84-99.
73. Wonnacott S, Soliakow L, Wilkie G, Redfern P, Marshall D. Presynaptic nicotinic acetylcholine receptors in the brain. Drug Devel Res 1996; 38:149-159.
74. Sivilotti L, Colquhoun D. Acetylcholine receptors: too many channels, too few functions. Science 1995; 269:1681-1682.
75. Hunter BE, de Fiebre CM, Papke RL, Kem WR, Meyer EM. A novel nicotinic agonist facilitates long-term potentiation in the rat hippocampus. Neurosci Lett 1994; 168:130-134.
76. Morales MA, Bachoo M, Collier B, Polosa C. Pre- and post-synaptic components of long-term potentiation in the superior cervical ganglion of the cat. J Neurophysiol 1994; 72:819-824.

77. Steinlein O, Mulley J, Propping P, Wallace R, Phillips H, Sutherland G, Schafer J, Berkovic S. A missense mutation in the neuronal nicotinic acetylcholine receptor α4 subunit is associated with autosomal dominant nocturnal frontal lobe epilepsy. Nature Genetics 1995; 11:201-203.
78. Picciotto MR, Zoli M, Lena C, Bessis A, Lallemand Y, Le Novère N, Vincent P, Pich EM, Brulet P, Changeux J-P. Abnormal avoidance learning in mice lacking functional high-affinity nicotine receptor in the brain. Nature 1995; 374:65-67.
79. Freeman JA. Possible regulatory function of acetylcholine receptor in maintenance of retinotectal synapses. Nature 1977; 269:218-222.
80. Chan J, Quik M. A role for the neuronal nicotinic α-bungarotoxin receptor in neurite outgrowth in PC12 cells. Neurosci 1993; 56:441-451.
81. Pugh PC, Berg DK. Neuronal acetylcholine receptors that bind α-bungarotoxin mediate neurite retraction in a calcium-dependent manner. J Neurosci 1994; 14:889-896.
82. Zheng JQ, Felder M, Connor JA, Poo M-m. Turning of nerve growth cones induced by neurotransmitters. Nature 1994; 368:140-144.
83. Bencherif M, Lukas RJ. Cytochalasin modulation of nicotinic cholinergic receptor expression and muscarinic receptor function in human TE671/RD cells: A possible functional role of the cytoskeleton. J Neurochem 1993; 61:852-864.
84. Lukas RJ. Diversity and patterns of regulation of nicotinic receptor subtypes. Annals NY Acad Sci 1995; 757:153-168.
85. Hory-Lee F, Frank E. The nicotinic blocking agents d-tubocurarine and α-bungarotoxin save motoneurons from naturally occurring death in the absence of neuromuscular blockade. J Neurosci 1995; 15:6453-6460.
86. Renshaw GMC, Dyson SE. α-BTX lowers metabolism during the arrest of motoneurone apoptosis. Neuroreport 1995; 6:284-288.
87. Treinin M, Chalfie M. A mutated acetylcholine receptor subunit causes neuronal degeneration in *C. elegans*. Neuron 1995; 14:871-877.
88. Akaike A, Tamura Y, Yokaota I, Shimohama S, Kimura J. Nicotine-induced protection of cultured cortical neurons against N-methyl-D-aspartate receptor-mediated glutamate cytotoxicity. Brain Res 1994; 644:181-187.
89. Martin E, Panickar K, King M, Deyrup M, Hunter B, Wang G, Meyer E. Cytoprotective actions of 2,4-dimethoxybenzylidene anabaseine in differentiated PC12 cells and septal cholinergic neurons. Drug Devel Res 1994; 31:135-141.
90. Janson AM, Møller A, Hedlund PB, von Euler G, Fuxe K. Nicotine and animal models of Parkinson's disease. In: Clarke PBS, Quik M, Adlkofer F, Thurau K, eds. Effects of Nicotine on Biological Systems II. Basel: Birkhäuser Verlag, 1995:127-135.
91. El-Dada MD, Quik M. Involvement of nitric oxide in nicotinic receptor-mediated myopathy. J Pharm Exper Thera 1997; 281:1463-1470.

92. Sharp B, Matta S. Activation of the hypothalamic-pituitary-adrenal axis by nicotine: Neurochemical and neuroanatomical substrates. In: Clarke PBS, Quik M, Adlkofer F, Thurau K, eds. Effects of Nicotine on Biological Systems II. Basel: Birkhäuser Verlag, 1995:159-166.
93. Ke L, Lukas RJ. Effects of steroid exposure on ligand binding and functional activities of diverse nicotinic acetylcholine receptor subtypes. J Neurochem 1996; 67:1100-1112.
94. Lukas RJ, Eisenhour CM. Interactions between tachykinins and diverse, human nicotinic acetylcholine receptor subtypes. Neurochem Res 1996; 21:1245-1257.
95. Bencherif M, Eisenhour CM, Prince RJ, Lippiello PM, Lukas RJ. The "calcium antagonist" TMB-8 [3,4,5-trimethoxy benzoic acid 8-(diethylamino)octyl ester] is a potent, noncompetitive, functional antagonist at diverse nicotinic acetylcholine receptor subtypes. J Pharm Exper Thera 1995; 275:1418-1426.
96. Musgrave IF, Krautwurst D, Schultz G. Drugs with high affinity for imidazoline receptors inhibit activation of nicotinic acetylcholine receptors. Mol Pharm 1997; (in press).
97. Donnelly-Roberts DL, Arneric SP, Sullivan JP. Functional modulation of human "ganglionic-like" neuronal nicotinic acetylcholine receptors (AChRs) by L-type calcium channel antagonists. Biochem Biophys Res Comm 1995; 213:657-662.
98. US DHHS. The Health Consequences of Smoking: Nicotine Addiction, A Report of the Surgeon General DHHS (CDC) 88-8406, Washington, D.C.: US Gov Print Off, 1988:618.
99. Gray JA, Mitchell SN, Joseph MH, Grogoryan GA, Dawe S, Hodges H. Neurochemical mechanisms mediating the behavioral and cognitive effects of nicotine. Drug Devel Res 1994; 31:3-17.
100. Henningfield JE, Schuh LM, Heishman SJ. Pharmacological determinants of cigarette smoking. In: Clarke PBS, Quik M, Adlkofer F, Thurau K, eds. Effects of Nicotine on Biological Systems II. Basel: Birkhäuser Verlag, 1995:127-135.
101. Warburton DM. The functional conception of nicotine use. In: Clarke PBS, Quik M, Adlkofer F, Thurau K, eds. Effects of Nicotine on Biological Systems II. Basel: Birkhäuser Verlag, 1995:257-264.
102. Peto R, Lopez A, Boreham J, Thun M, Heath C. Mortality from tobacco in developed countries: Indirect estimation from national vital statistics. Lancet 1992; 339:1268-1278.
103. Peng X, Anand R, Whiting P, Lindstrom J. Nicotine-induced upregulation of neuronal nicotinic receptors results from a decrease in the rate of turnover. Mol Pharm 1994; 46:523-530.
104. Barrantes GE, Rogers AT, Lindstrom J, Wonnacott S. α-Bungarotoxin binding sites in rat hippocampal and cortical cultures: Initial characterization, colocalisation with α7 subunits and up-regulation by chronic nicotine treatment. Brain Res 1995; 672:228-236.

105. Bencherif M, Fowler K, Lukas RJ, Lippiello PM. Mechanisms of upregulation of neuronal nicotinic acetylcholine receptors in clonal cell lines and primary cultures of fetal rat brain. J Pharm Exper Thera 1995; 275:987-994.
106. Lukas RJ, Ke L, Bencherif M, Eisenhour CM. Regulation by nicotine of its own receptors. Dev Drug Res 1996; 38:136-148.
107. Collins AC, Marks MJ. Are nicotinic acetylcholine receptors activated or inhibited following chronic nicotine treatment? Drug Devel Res 1996; 38:231-242.
108. Zhang X, Gong Z, Helstrom-Lindahl E, Nordberg A. Regulation of α4β2 nicotinic acetylcholine receptors in M10 cells following treatment with nicotinic agents. Neuroreport 1994; 6:313-317.
109. Lukas RJ. Effects of chronic nicotinic logand exposure on functional activity of nicotinic acetylcholine receptors expressed by cells of the PC12 rat pheochromocytoma or the TE671/RD human clonal line. J Neurochem 1991; 56:134-1145.
110. Hsu Y-N, Amin J, Weiss DS, Wecker L. Sustained nicotine exposure differentially affects α3β2 and α4β2 neuronal nicotinic receptors expressed in *Xenopus* oocytes. J Neurochem 1996; 66:667-675.
111. Rosecrans JA, Karan LD, James JR. Nicotine as a discriminative stimulus: individual variability to acute tolerance and the role of receptor desensitization. In: Clarke PBS, Quik M, Adlkofer F, Thurau K, eds. Effects of Nicotine on Biological Systems II. Basel: Birkhäuser Verlag, 1995:219-223.
112. Garcia-Guzman M, Sala F, Sala S, Campos-Caro A, Criado M. Role of two acetylcholine receptor subunit domains in homomer formation and intersubunit recognition, as revealed by α3 and α7 subunit chimeras. Biochem 1994; 33:15198-15203.
113. Albuquerque EX, Pereira EFR, Bonfante-Cabarcas R, Marchioro M, Matsubayashi H, Alkondon M, Maelicke A. Nicotinic acetylcholine receptors on hippocampal neurons: Cell compartment-specific expression and modulatory control of channel activity. In: J Klein, K Löffelholz, eds. Cholinergic Mechanisms: From Molecular Biology to Clinical Significance. Vol. 109, Amsterdam, Elsevier, Progress Brain Res, 1996: 111-124.
114. DeFiebre C, Meyer E, Henry J, Muraskin SWK, Papke R. Characterization of a series of anabaseine-derived compounds reveals that the 3-(4)-dimethylaminocinnamylidine derivative is a selective agonist at neuronal nicotinic α7/^{125}I- receptor subtypes. Mol Pharm 1995; 47:164-171.
115. Lippiello PM, Bencherif M, Caldwell WS, Arrington SR, Fowler KW, Lovette ME, Reeves LK. Metanicotine: a nicotinic agonist with central nervous system selectivity—in vitro and in vivo characterization. Drug Devel Res 1996; 38:169-176.
116. Brioni JD, Decker MW, Sullivan JP, Arneric SP. The pharmacology of (–)-nicotine and novel cholinergic channel modulators. Adv Pharm 1997; 37:153-214.

117. Russell MAH, Stapleton JA, Feyerabend C. Nicotine boost per cigarette as the controlling factor of intake regulation by smokers. In: Clarke PBS, Quik M, Adlkofer F, Thurau K, eds. Effects of Nicotine on Biological Systems II. Basel: Birkhäuser Verlag, 1995:233-238.
118. Silver A, Shytle R, Philipp M, Sandberg P. Transdermal nicotine in Tourette's syndrome. In: Clarke PBS, Quik M, Adlkofer F, Thurau K, eds. Effects of Nicotine on Biological Systems II. Basel: Birkhäuser Verlag, 1995:293-299.
119. Newhouse P, Potter A, Corwin J. Effects of nicotinic cholinergic agents on cognitive functioning in Alzheimer's and Parkinson's disease. Drug Devel Res 1996; 38:278-289.
120. Freedman R, Leonard S, Alder L, Bickford P, Byerley W, Coon H, Miller C, Luntz-Leybman V, Myles-Worsley M, Nagamoto H, Rose G, Stevens K, Waldo M. Nicotinic receptors and the pathophysiology of schizophrenia. In: Clarke PBS, Quik M, Adlkofer F, Thurau K, eds. Effects of Nicotine on Biological Systems II. Basel: Birkhäuser Verlag, 1995:307-312.
121. Levin ED, Conners CK, Sparrow E, Hinton S, Meck W, Rose JE, Ernhardt D, March J. Nicotine effects on adults with attention deficit/hyperactivity disorder. Psychopharm 1996; 123:55-63.
122. Thomas GAO, Rhodes J. Relationship between smoking, nicotine and ulcerative colitis. In: Clarke PBS, Quik M, Adlkofer F, Thurau K, eds. Effects of Nicotine on Biological Systems II. Basel: Birkhäuser Verlag, 1995:287-291.
123. Lange KW, Wells FR, Senner P, Marsden CD. Altered muscarinic and nicotinic receptor densities in cortical and subcortical regions in Parkinson's disease. J Neurochem 1993; 60:197-203.
124. Baron J. The epidemiology of cigarette smoking and Parkinson's disease. In: Clarke PBS, Quik M, Adlkofer F, Thurau K, eds. Effects of Nicotine on Biological Systems II. Basel: Birkhäuser Verlag, 1995:313-319.
125. Lee PN. Smoking and Alzheimer's disease: A review of the epidemiological evidence. Neuroepidem 1994; 13:131-144.
126. Poirier J, Aubert I, Bertrand P, Quirion R, Gauthier S, Nalbantoglu J. Apolipoprotein E4 and cholinergic dysfunction in AD. In: Giacobini E, Becker R, eds. Alzheimer's Disease: Therapeutic Strategies. Boston: Birkhauser, 1994:72-76.
127. Daly J. The chemistry of poisons in amphibian skin. Proc Natl Acad Sci USA 1995; 92:9-13.
128. Houghtling RA, Dávila-Garcia MI, Kellar KJ. Characterization of (±)-[^{3}H]epibatidine binding to nicotinic cholinergic receptors in rat and human brain. Molec Pharm 1995; 48:280-287.
129. Lindstrom J, Shelton GD, Fuji Y. Myasthenia gravis. Adv Immunol 1988; 42:233-284.
130. Engel A. Myasthenic syndromes. In: Engel A, Franzini-Armstrong C, eds. Myology, 2nd ed., Vol. 2. New York: McGraw-Hill, 1994: 1798-1835.

131. Ohno K, Hutchinson DO, Milone M, Brengman JM, Bouzat C, Sine SM, Engel AG. Congenital myasthenic syndrome caused by prolonged acetylcholine receptor channel openings due to a mutation in the M2 domain of the ε subunit. Proc Natl Acad Sci USA 1995; 92:758-762.
132. Stitzel JA, Robinson SF, Marks MJ, Collins AC. Differences in response to nicotine are determined by genetic factors. In: Clarke PBS, Quik M, Adlkofer F, Thurau K, eds. Effects of Nicotine on Biological Systems II. Basel: Birkhäuser Verlag, 1995:279-284.

CHAPTER 8

Molecular Pathology of the Nicotinic Acetylcholine Receptor

Francisco J. Barrantes

Introduction

Ion channels and receptors play a central role in a variety of cell functions. Accordingly, they can be affected by a variety of pathological conditions leading to abnormal cell function, either through an inherited condition or in an acquired form. The nicotinic acetylcholine receptor (AChR) is no exception, and is known to be the target of several inherited and acquired diseases.

Our current knowledge of the structure and function of the AChR, obtained through interdisciplinary approaches in the last two decades, has recently reached a stage of maturity enabling the description of various pathological conditions affecting this protein with an unprecedented level of detail. The possible exploitation of new genetic tools in the diagnosis of novel receptor pathologies is proposed in this chapter, particularly genome screening of large sequence databases to identify the animal complement of human genes and new members of gene families, to discover novel phenotypes of medical interest, and to unravel mutations of the AChR and other ligand-gated channels leading to heritable diseases. Of particular relevance for the future are those diseases associated with mental and behavioral disorders that have so far escaped elucidation.

Pathologies of the AChR and other ion channels can initially be dichotomized into direct, involving a mutation in the gene coding for some critical part of the protein, or indirect, comprising

The Nicotinic Acetylcholine Receptor: Current Views and Future Trends,
edited by Francisco J. Barrantes.

autoimmune disease or defective coding of a key nonreceptor protein acting as regulator or modulator of the receptor/ion channel proper (see Table 8.1 and ref. 1).

From a different perspective, diseases affecting the AChR and other channel proteins can be classified as genetic and somatic. Among the pathologies directly affecting the AChR protein, we may initially consider single or multiple mutations affecting one or more subunits or domains of the oligomeric protein. These mutations can either be silent and pass unnoticed, or manifest themselves in physiopathological terms. More severe genetic alterations involving missense or nonsense mutations may result in partial deletion, truncation, or translocation of AChR domains, but still permit receptor assembly and trafficking to the cell surface. Defective subunit assembly of the AChR, incompatible with its subsequent targeting to the plasmalemma, may result in similar but more severe defects, probably involving either deletion of entire regions of the subunits or alteration of critical domains involved in intersubunit contacts and/or assembly. Some spontaneous mutations of receptor-ion channel proteins of this latter type are likely to be lethal, but other alterations allow survival of the diseased cell and/or the individual. The latter we qualify as ion channel hereditary diseases. The recent knockout experiments with genetically engineered pathological receptor subunits make apparent the extraordinary capacity of transgenic animals to compensate for the defective genes (embryonic stem cell ("knockout") animal models are single-gene disruption mutants that enable the evaluation of the null-phenotype of a gene in the whole animal). Indirect channel pathologies encompass all of the abnormal conditions summarized in Table 8.1.

In the case of genetic diseases affecting ion channels and receptors, two basic strategies have been employed for the characterization of the pathology at the molecular level. The first approach is based on the fact that many inherited diseases of dominant nature are caused by mutations in the gene coding for the ion channel-receptor protein, leading to altered function of the resulting protein. Thus, assays of the receptor functional properties may result in the identification of the pathology at the genomic, molecular level. Once the faulty gene has been identified, characterization of the mutations follows. This strategy has been successfully used for the identification of the diseased genes responsible for hyperkalemic periodic paralysis and for malignant hyperthermia, in which cases Na^+ and

Table 8.1. Direct and indirect pathologies affecting the AChR

Direct AChR pathologies

(a) single or multiple dominant (or more rarely recessive) nonsense or missense mutations affecting one or more AChR subunits (silent / apparent); these include most AChR ion channel congenital, hereditary syndromes that slow down (e.g., SCCMS) or accelerate (fast channel syndrome) response to ACh (both of which will impair synaptic transmission;

(b) defective AChR assembly resulting from mutations in regions involved in AChR oligomerization; other subunits may replace (i.e., "phenotypic rescue") for impaired subunit (e.g., some SCCMS);

(c) more severe genetic alterations involving partial deletion / truncation / translocation of entire AChR subunits or domains, resulting in abnormal AChR oligomer formation, some of which may result in total impairment of cell surface expression of the AChR protein.

Indirect AChR pathologies

(a) abnormalities in extrinsic regulatory mechanisms (phosphorylation, acylation, glycosylation, etc.) required for AChR expression, trafficking, etc.; some of these diseases may result from altered regulation of gene cluster(s) by a locus control region;

(b) defective regulation of AChR metabolic stability at the cell surface, related to aggregation and/or faulty relationship with nonreceptor proteins;

(c) presence of substances acting as channel blockers (e.g., toxic substances);

(d) autoimmune attack on channel protein or associated molecules;

(d) pathologies of lipid metabolic pathways resulting in secondary AChR pathology.

Ca^+ channels are affected, respectively. The second approach is more akin to genetics, and is based on pedigree linkage studies, that is, epidemiological and genetic surveys of families affected by a given disease, using genetic markers of known chromosomal location, in an attempt to identify responsible genes. Genetic mapping of the target chromosome is used next to refine the positioning of the diseased gene. Once the locus is characterized, cloning, exploration for open reading frames, and search for sequence abnormalities is undertaken over large DNA stretches. Identification of the genes responsible for the Cl^- channel pathology associated with cystic fibrosis and other recessive genetic diseases are examples of this type of strategy.[2]

Neurotransmitter receptor dysfunction in psychiatric disorders, especially those comprising alterations of affection and/or cognition, is being increasingly studied by the pharmaceutical industry in view of its enormous market potential. The possibility of developing appropriate therapeutic agents for the treatment of anxiety, alcoholism, schizophrenia, aggressive disorders, suicidal behavior, migraine, and other neuropsychiatric disorders is the engine of much of the research carried out in this field. Other areas of Medicine are becoming increasingly aware of the prophylactic and therapeutic potentialities of understanding receptor and ion channel pathology. Thus, pharmaceutical agents that act on ion channels have been sought in the treatment of aberrant electrical excitability, as in epileptic disorders or cardiac arrhythmias, as the target of local and general anesthetics, and in modulating vascular tone and epithelial function. There is also increasing interest in ion channels and neurotransmitter receptors as targets for treatment of metabolic diseases and immune system modulation.

One of the prerequisites for development of useful therapeutic agents is the understanding of ion channel or neurotransmitter receptor structure and function. This requires, in turn, the combined effort of electrophysiology, molecular biology, pharmacology, natural product (organic), and protein chemistry, and structural biology with a focus on the pathogenesis of the diseased conditions. The complementation will certainly bring not only novel approaches to drug discovery and therapy but, interestingly enough, will also result in a refinement of our current knowledge of the structure and function of ion channels and neurotransmitter receptors at the molecular level. In this chapter I have chosen some of the emerging pathologies of the AChR whose knowledge has been made possible through recent advances in the molecular characterization of the AChR molecule. The reader is referred to a volume of this series[3] and reviews (e.g., ref. 4) for a comprehensive treatment and extensive literature coverage of myasthenia gravis (MG), the most thoroughly studied disease affecting nicotinic AChR and several other diseases involving this cell surface receptor. In this chapter I shall only cover some selected aspects of this pathology before discussing congenital myasthenic syndromes (CMS) and other channel pathologies beginning to be characterized at the molecular level, mostly affecting the kinetics of the AChR channel[5-8] and for which I propose the term "molecular diskynesias." I shall also review some possible therapeutic implications of recent experimental work from our laboratory. Finally, I shall analyze future trends in the discovery of new AChR diseases. Some rarer diseases in which the AChR may be affected, such as adjuvant induced polyarthritis,[9] lung cancer,[10] or lower motor neuron disease have been reviewed.[5] The important disorders associated with nicotine dependence are treated by Ronald Lukas in chapter 7 of this volume.

Myasthenia Gravis

MG, the most common acquired somatic disease involving the AChR, is an excellent example in which a disease can be causally linked with a receptor pathology. It is an autoantibody-mediated disorder in which the target of the antibodies is the AChR protein (see Fig. 8.1). As a consequence of the autoimmune reaction a marked decrease in the number of AChR molecules occurs, and neuromuscular transmission is impaired, with the resulting clinical manifestation of severe muscle weakness and fatigability. Electrophysiologically, smaller postsynaptic currents are observed in MG.

Autoimmune $CD4^+$ T helper cells recognize a large number of AChR epitopes in association with major histocompatibility complex (MHC) class II molecules on the surface of antigen-presenting cells.[12] Thus, during the fully developed anti-AChR response in MG the receptor protein proper is the molecular target of the $CD4^+$ T-cells. Anti-AChR antibodies in MG patients are high-affinity IgGs, whose production requires the intervention of specific $CD4^+$ T-helper (Th) cells. MG signs can be transferred to healthy animals or reproduced in vitro by administering such IgG fraction from the serum of MG patients, or poly-/monoclonal antibodies from animals suffering from experimental autoimmune MG (EAMG). Experimentally produced monoclonal antibodies have been shown to cause the channel to make kinetic transitions leading to desensitization rather than activation. Most anti-AChR antibodies in the sera of laboratory animals that develop EAMG recognize a region in the extracellular domain of the AChR α-subunit called the main immunogenic region (MIR,[13]). Anti-MIR antibodies are responsible for the capacity of the sera of MG patients to cause AChR loss in cell cultures. The sequence in the AChR α-subunit that contains the MIR (α65-80) exhibits high sequence homology with a region of the U1 small nuclear ribonucleoprotein, an important autoantigen in systemic lupus erythematosus, and of a protein found in several retroviruses,[14] a fact which has led to speculation on possible kindredness of MG with lupus and a putative viral component in the etiology of MG, respectively.

Fig. 8.1. Synaptic view of the AChR with monoclonal antibody molecules bound to its main immunogenic region (MIR). The three-dimensional surface view of an assembly of AChR molecules (doughnut shaped objects) embedded in the membrane in complex with anti-MIR mAbs (oblong bodies), as observed in about two-thirds of patients with the autoimmune disease MG. The mAb fragments are monovalent, yet they appear to bind to two adjacent AChR molecules. In fact, individual mAb fragments are not resolved in the micrographs. The illustration, reconstructed from cryoelectron micrographs, was kindly provided by Dr. Nigel Unwin, MRC Laboratory of Molecular Biology, Cambridge, England (cover picture of Neuron 1995; 15(2), reprinted with permission). For details see ref. 11.

Neonatal weakness occurs in less than 10% of pregnancies of MG patients: the autoantibodies cross the placental barrier (reviewed in ref. 15). There is also a fetal form of MG in which autoantibodies against the fetal, ε-type AChR are observed, together with congenital abnormalities (reviewed in ref. 16). Polyclonal anti-AChR autoantibodies have been implicated in AChR loss, and are observed in the serum in 85% of the patients,[17] although there is no correlation between antibody titers and severity of the disease. This is thought to reflect the fact that only some subpopulations of anti-AChR are pathogenetic, probably because of their higher capacity to stimulate antigenic modulation or complement activation. Patients with clinical signs of acquired MG but without detectable levels of anti-AChR autoantibodies are denominated "seronegative MG." Plasma from these seronegative MG individuals reduces AChR function in muscle cells in culture. Babies born of MG mothers may have transient neonatal myasthenia. Most of the MG syndromes are apparently due to a T lymphocyte-dependent serum autoantibody against AChR.[18]

The detailed pathogenesis of MG is still unknown. The AChR in muscle, an AChR-like protein elsewhere, or a cross-reacting protein have been implicated. The thymus has been proposed as the arena for the anti-AChR sensitization of relevant cells. The thymus in MG patients may be pathological: it is often hypertrophic, may have a thymoma and may contain anti-AChR T and B cells. Conversely, 30-40% of patients with thymomas also suffer from MG. Thymectomy is still performed in some MG patients and is symptomatically beneficial (reviewed in ref. 19).

Among the symptomatic treatments of choice in the therapeutic arsenal against MG, immunosuppressive drugs have received preferential attention. Unfortunately, these drugs produce a generalized suppression of the immune system. Ideally, in the treatment of MG one would like to selectively suppress the exacerbated response to the AChR as an antigen without adversely affecting the rest of the immune system. Oral administration of antigens has been used in the treatment of cell-mediated experimental autoimmune diseases. *Torpedo* AChR was administered to rat as antigen in EAMG, an experimental model of MG.[20] Cellular responses to feeding with native, intact AChR, as measured by interleukin production and lymphocyte proliferation, were found to be markedly inhibited, suggesting that oral therapy is beneficial in EAMG and may be useful in MG patients; however, molecules with less immunogenic potential than the intact AChR molecule may be better suited for this purpose.[20]

Plasma exchange is effective in temporarily relieving the symptoms of MG, but several drawbacks are associated with this therapeutic strategy. Repeated plasmapheresis increases the risk of contracting transmitted diseases, and other strategies have been sought to ameliorate the symptoms of MG. Removal of anti-AChR antibodies by protein A immunoadsorption has been proposed.[21] Since staphylococcal protein A is a strong immunoadsorbant for immunoglobulins that interacts negligibly with other plasma proteins, the anti-AChR IgG immunoadsorption with protein A resulted in a 70-80% removal of these immunoglobulins with a concomitant clinical improvement in the patients.

Another substitute approach for immunosuppressive therapy of MG has been put forward by Araga et al.[22] An idiotype is an antigenic determinant associated with the antigen-binding site of an antibody molecule. Since production of antibodies against this site ("anti-idiotypic antibody") should in principle be a highly specific way of treating MG by neutralizing anti-AChR antibodies, these authors attempted the active production of anti-idiotypic antibodies by immunization with a peptide (termed (α67-76) encoded by RNA complementary to the *Torpedo* AChR α-subunit containing the MIR. The anti-idiotypic antibody-inducing ability of this peptide was demonstrated. The experimental form of the disease, EAMG, in rats challenged with *Torpedo* AChR, was milder with respect to controls.

Several muscle disorders affect extremity and ocular muscles differently. Polymyositis, myotomas, peripheral paralysis and most dystrophies rarely produce ocular signs. MG patients exhibit a prominent weakness of the extraocular muscles (EOM) in 90% of the cases. Furthermore, EOM weakness often appears in early stages of the disease. In addition, 15% of all MG patients will manifest only ocular signs. It has been argued that the high firing frequency of the motor units innervating EOM may increase their susceptibility to fatigue.[23] In fact, 80% of EOM fibers have a neuromuscular junction, the remaining receiving multiple synapses per fiber. Single-terminal fibers in the EOM are equivalent to the fast-twitch extremity muscle fibers, producing a synchronized contraction and a propagated action potential in response to a single nerve stimulus. But EOM responses are twice as fast as those of the extremity fast-twitch fibers, enabling them to operate at high firing frequencies. Multiterminal EOM fibers, on the other hand, do not generate action potentials, and are mostly associated with slow synaptic current kinetics which cause a uniform depolarization of tonic fibers, thus maintaining a graded, slow muscle contraction in response to nerve stimulation. The higher propensity of fast-twitch EOM to be affected in MG syndromes may be a consequence not only of these two anatomical and physiological properties, but also of the lower density of AChRs at EOM twitch synapses.[24] Higher motor unit frequencies coupled to a smaller number of AChRs could easily lead to exhaustion of neuromuscular transmission and thus make twitch fibers more vulnerable to fatigue than extremity muscles. But of the two types of EOMs, the multiterminal fibers, similar to nonmammalian tonic fibers, would be even more fatigue-prone.[25]

Soon after birth, the slow fetal type ($\alpha_2\beta\gamma\delta$)-AChR is replaced by the fast, adult ($\alpha_2\beta\varepsilon\delta$)-type AChR.[26] Serum from myasthenic patients contains antibodies that react selectively with fetal-type junctional AChR. Twitch fibers contain both adult, ε-type and embryonic, γ-type AChR channels with a longer mean open time (that is, a longer average duration of the AChR channel in the open state), thus enabling these fibers to better respond to repeated or prolonged nerve stimulation. EOM was subsequently found to express embryonic, γ-type AChR, in contrast to other striated skeletal muscle in the adult.[27] Kaminski et al[28] have investigated whether the γ-subunit is also expressed in levator palpebrae superioris, a muscle affected in MG but known not to have multiterminal fibers. No transcripts of the fetal type were found, indicating that the susceptibility of this muscle to MG is not associated with γ-subunit fetal-type AChR expression.

Although the predominant action of corticoid hormones is exerted on the synthesis of a restricted number of specific mRNA species in target cells, direct action of steroids on ion channels has also been reported. Glucocorticoids find therapeutic use in conjunction with AChE inhibitors for the treatment of the MG by virtue of their immunosuppressive action, but the clinical symptoms of MG often deteriorate during the initial phase of glucocorticoid treatment, when relatively high doses are used. This prompted us to study the acute effect of glucocorticoids on the AChR at the single-channel level.[29-30] First we found that exposure of BC3H-1 cells (expressing endogenous embryonic-type AChR) to hydrocortisone induced a dose-dependent reduction in the channel open time and burst duration and an increase in the closed time, with no changes in channel amplitude.[29-30] At 1 mM hydrocortisone, the AChR channel lifetime was about sixfold shorter than that of the control. Similar effects were observed with 11-desoxycortisone, thus suggesting that the oxygen function at position 11 is not required for channel modification. In another series of experiments, we found that another synthetic glucocorticoid, dexamethasone, induced: (i) a dose-dependent shortening of the channel mean open time; (ii) grouping of single-channel openings into bursts and (iii) flickering substructure of the bursts of openings.[30] These acute effects could be described most economically by the linear kinetic scheme below:

$$A_2R^* \underset{b}{\overset{f[B]}{\rightleftharpoons}} A_2B \qquad (1)$$

The above scheme is an "arm" of the one presented in the introduction, extending the open biliganded state A_2R^* to A_2B, a blocked state of the AChR channel in the presence of dexamethasone. f and b are the forward and backward rate constants for channel blocking, respectively. Considering only one open-blocked state, the mean open time would decrease from 1/a to 1/ $(\alpha + f[B])$. From the data, a value of f of $7.3 \times 10^5\ M^{-1}\ s^{-1}$ was calculated.[29-30]

For the simple sequential blocking model (scheme 1 above), the burst duration is expected to increase as a function of blocker concentration because blocking events prolong the time that channels spend open before closing. In dexamethasone-modified channels the burst duration decreased as a function of ligand concentration, a finding inconsistent with the idea that dexamethasone acts as a simple open channel blocker. It is possible that the AChR is able to undergo a conformational transition from the open-blocked (A_2R^*B) to a closed-blocked state (A_2B) while interacting with—and being blocked by—the steroid, thus making Scheme 2 below (an extension of Scheme 1 above) more appropriate to describe the action of the drug:

$$A_2R \underset{\beta}{\overset{\alpha}{\rightleftharpoons}} A_2R^* \underset{b}{\overset{f[B]}{\rightleftharpoons}} A_2R^*B \underset{\alpha'}{\overset{\beta'}{\rightleftharpoons}} A_2B \qquad (2)$$

The impairment of the AChR channel function by dexamethasone thus provides an explanation for the worsening of myasthenic symptoms before the immunosuppressive effects of the steroid make themselves apparent. More recently we have extended our initial observations to the glucocorticoid hydrocortisone.[31]

Congenital Myasthenic Syndromes (CMS) and Other "Molecular Dyskinesias" of the AChR Channel

There are several rare, genetically heterogeneous inherited congenital myasthenic syndromes that evolve with the typical clinical symptoms of MG, and the characteristic morphological and electrophysiological signs of the conventional myasthenic disease, but without detectable autoantibodies against the AChR. These congenital disorders of neuromuscular transmission differ from MG and from the Lambert-Eaton myasthenic syndrome in that there is no autoimmune involvement. Vincent et al[6] have classified these syndromes into autosomal dominant—or sporadic—and recessive, as shown in Table 8.2.

These syndromes are usually observed at birth, and may progress slowly without marked variations in their severity. Histological findings comprise presynaptic alterations that affect neurotransmitter release, acetylcholinesterase deficiency, and postsynaptic abnormalities that involve a decreased number of AChRs and altered function of the channel (kinetic abnormalities).[45-46]

The so-called "slow channel congenital myasthenic syndromes" (SCCMS) is probably the first degenerative group of diseases to be characterized as resulting from congenital abnormalities of AChR function. This group of inherited progressive myopathies is characterized by weakness, especially of cervical and scapular muscles, eventually reaching atrophy, degenerative changes of the NMJ, and as recently identified, clear electrophysiological evidence of abnormal AChR channel function. The onset of the disease can occur after birth, or later in life. Family members can also manifest some of the electromyographic signs of the disease, such as the characteristic repetitive response to a single nerve stimulus observed in some afflicted patients, but otherwise be asymptomatic. Since the decay time of the miniature end-plate potentials at the neuromuscular junction

Table 8.2. Molecular anatomy of some diseases affecting the AChR protein

Disease	Zone or Residues Involved / Mutated	Functional Consequences	Reference
Myasthenia gravis	α_1 MIR in M1 (extracellular) residues 67-76 in α1	impaired neuromuscular transmission	reviewed in Conti-Fine et al[4]
CMS			
A. autosomal dominant missense mutations			
Slow-channel syndromes (SCCMS)	α_1 Gly153Ser (extracellular; binding domain?)	Higher ACh affinity; longer channels: reduced k_{-2} (agonist dissociation rate); increased openings per burst	Sine et al[32]
	α_1 Val156Met (extracellular; binding domain?)	stabilization of the openstate: longer openings	Croxen et al[8] Newland et al[33]
	α_1 Asn217Lys in M1 (extracellular)	longer channel openings:slower channel closure due to slower ACh dissociation; increased affinity for ACh; enhanced desensitization	Engel et al[34] Wang et al[48]
	α_1 Thr251Ile		Croxen et al[8]
	α_1 Thr254Ile (channel; 3 residues C-term from Leu ring)		Croxen et al[8]
	α_1 Ser269Ile (extracellular loop between M2/M3)	prolonged openings	Newland et al[33]
	β_1 Val262Met in M2) (channel region)	longer, leaky channels: slower channel closure; increased affinity for ACh	Croxen et al[8] Gomez et al[35]
	β_1 Val266Met in M2 (channel, 4 residues from leucine ring)	longer, leaky channels: slower channel closure; increased affinity for ACh	Engel et al[34]
	δ Gln267Glu in M2 (channel, but not lumenal)	rare benign polymorphism	Engel et al[34]

Table 8.2 (continued).

Disease	**Zone or Residues Involved / Mutated**	**Functional Consequences**	**Reference**
	ε Arg147Leu	altered assembly of adult-type AChR (γ-type predominates)	Ohno et al[47]
	ε Pro245Leu (C-term in M1)	slower rate of channel closure; reduced surface expression of AChR	Ohno et al[47]
	ε Thr264Pro in M2 (channel)	excessively prolonged channel open dwell time	Ohno et al[37]
	ε Leu269Phe in M2 (channel; 8 residues from leucine ring)	slower channel closure, increased ACh affinity, enhanced desensitization	Engel et al[34]
	ε Arg311Trp between C-term of M3 and cytoplasmic loop	reduced open channel intervals; reduced surface expression of AChR	Ohno et al[47]
B. Recessive			
AChR deficiency	reduced AChR number	reduced m.e.p.p.; improvement with AChE inhibitors	Vincent et al[16]
Fast channel congenital syndrome	ε Pro121Leu in M1 (extracellular)	reduced ACh affinity; infrequent single-channel events; reduced opening rate β; briefer channel openings	Ohno et al[38]
Nonmyopathic congenital myasthenic syndrome	substitution of ε by γ missense codons in ε	severe AChR deficiency; insertion of incorrect nucleotides in genes	Engel et al[39]
Epilepsy			
autosomal dominant nocturnal frontal lobe epilepsy (ADNFLE)	neuronal AChR α_4 subunit Ser248Phe in M2 (channel)	AChR channel altered conductance; nocturnal frontal lobe epilepsy	Steinlein et al[40]
		faster desensitization rate	Weiland et al[41] Figl et al[42]
benign familial neonatal convulsions (EBN1)	neuronal AChR α_4 subunit	nonsense mutation in the α4 subunit cosegregates with 20q-chromosome	Beck et al[43] Schubert et al[44]

is dictated by the lifetime of the activated A_2R^* state, m.e.p.p.s also exhibit abnormally prolonged decay times. This can result from several "microscopic" kinetic anomalies:

1) inherently abnormal prolongation of the biliganded open state, A_2R^* (see Scheme (3) below);
2) increase in the number of openings per burst;
3) a combination of the two.

$$2A+R \underset{k_{-1}}{\overset{k_1}{\rightleftharpoons}} A+AR \underset{k_{-2}}{\overset{k_2}{\rightleftharpoons}} A_2R \underset{\alpha}{\overset{\beta}{\rightleftharpoons}} A_2R^* \qquad (3)$$

Thus the term SCCMS alludes to the underlying kinetic anomaly at the molecular level: the AChR spends an abnormally long time in the open state. The prediction was made that such kinetic behavior is associated with mutations in the AChR protein.[46] Indeed, in the first of these syndromes to be described at the single-channel level,[37-38] a reduced number of AChRs was found, with two populations of mean open times, one of which displayed abnormally prolonged durations. Molecular genetic analysis of the AChR genes revealed a single base mutation at nucleotide 790 in exon 8 of the ε-subunit gene, which codes for the amino acid proline 264. Mutant AChR was constructed to mimic the endogenous mutation of proline for threonine at this position, located at the M2 transmembrane segment of the ε-subunit. At the single-channel level, the engineered mouse ε-type AChR homozygously expressed in HEK-293 cells displayed the same electrophysiological properties as the heterozygous mutated human AChR observed in biopsies of skeletal muscle from the patient.

The work of Ohno et al[37] raised the possibility that similar point mutations in sensitive areas of other ligand-gated channels may also exist in other neurological and psychiatric disorders. Indeed, other CMS have since been described,[7,8,34,35,39,47,48,] all having in common abnormal AChR channel opening episodes (see Table 8.2). Several of these pathogenic mutations in the M2 transmembrane segment of the AChR occur in the outer, synaptic region of the channel, or pore region proper (see Fig. 8.3). Engel et al[34] also point out that these mutations are positioned towards the channel lumen, some of them very close to the leucine ring,[49] and they involve introduction of a larger side chain than the original amino acid residue in the corresponding location. Mutations in channel lumen-facing amino acids may thus have in common pathologically slow channel closing rates, with the resulting prolonged opening episodes.

Other SCCMS reported to date also occur in AChR subunits other than ε (α, β). At the molecular level, the defects are found in the channel-lining transmembrane segment M2, and also in M1.[34,37-38] In some reports the mutations associated with a SCCMS are found in the extracellular domain of the α subunit.[32-33] Mutation αG153S is one such case (Table 8.2). It occurs in a region purported to contribute, lie or close, to the agonist-recognition site (Fig. 8.3), thus rationalizing the observation that the pathologically prolonged open time results in this case from the reduced agonist dissociation rate. The prolonged m.e.p.p.s observed in the different SCCMS result from prolonged bursts of openings, a consequence in turn of the abnormally long duration of single-channel openings (usually associated with a reduced rate of channel closure, α), an increased number of single-channel openings per burst, or a combination of the two. In a recent study, Gomez et al[35,50] have produced transgenic mouse lines carrying SCCMS. Nerve-evoked end-plate currents and m.e.p.c.s had prolonged decay times and their amplitudes were reduced by 33%. Transgenic mice were abnormally sensitive to the neuromuscular blocker curare.

Knowledge of the pathological alterations occurring in SCCMS at the molecular level has important consequences for therapeutic strategies, making it clear for instance that acetylcholinesterase inhibitors used in the treatment of MG should not be employed in SCCMS. Capitalizing on previous observations on the effect of steroids on AChR channel behavior, Cecilia Bouzat and I investigated the effect of synthetic glucocorticoid on the mouse model of one of the human SCCMS, $\alpha_2\beta\varepsilon_{T264P}\delta$. This mutation dramatically prolongs the open channel dwell time of the AChR.[37] Hydrocortisone was found to significantly reduce the abnormally long mean open time (ref. 31 and Fig. 8.2). Our results open the way for research on other therapeutic strategies using channel blockers devoid of the long-term effects of steroid administration.

The pathologically lengthy dwell times in the open channel state observed in the SCCMS appear to lead, in some cases, to a myopathy with loss of AChR owing to destruction of the junctional folds at the end-plate region. This morphological degeneration is most likely caused, in turn, by cationic overloading of the postsynaptic skeletal muscle cell. Interestingly, the *deg-3* (u662) mutation associated with

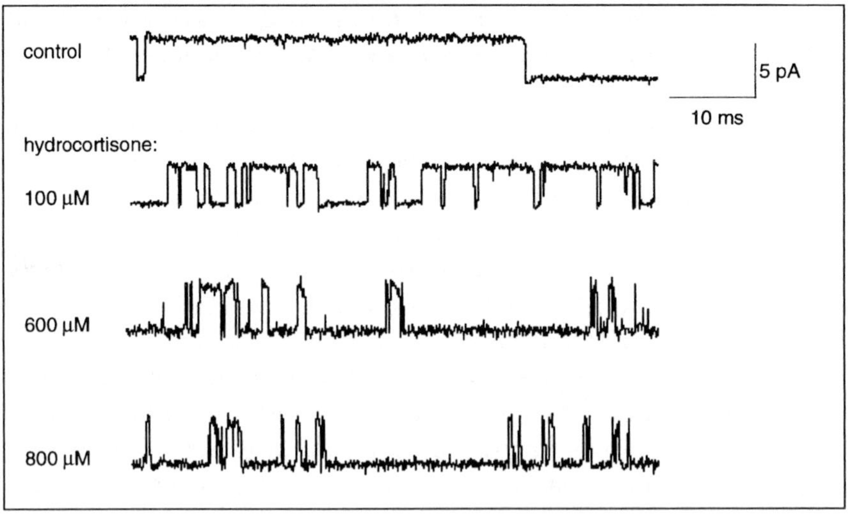

Fig. 8.2. Single-channel recordings showing the effect of increasing concentrations of hydrocortisone on the mean open time of mouse AChR carrying the $\alpha_2\beta\epsilon_{T264P}\delta$ mutation, found in humans with one of the slow channel congenital myasthenic syndromes (see Table 8.2 and ref. 37). See further details in ref. 31.

the degeneration of some neurons in *C. elegans* lies within the AChR αM2 transmembrane domain, and is associated with prolonged bursts of channel openings, resulting in exocytoxicity.[51]

It can be anticipated that future screening for mutations in the AChR subunit genes using two-armed analysis and single-stranded conformation polymorphism analysis of patients with myasthenic syndromes, in combination with patch-clamp analysis of "synthetic" mutants expressed in heterologous cellular systems, will enable the identification of more SCCMS variants. This screening may combine clinical, population genetics, molecular biology, and electrophysiological studies. DNA from clinically identified patients will be submitted to nucleotide sequencing. Regions possessing single-strand polymorphism, and the sequences of the four transmembrane domains of each AChR subunit in cases in which no polymorphism is found, will be identified. The functional correlate of any SCCMS clinical phenotype with mutations that is found will be verified by reproducing the human cDNA pathology with the corresponding AChR subunit cDNA mutation in mouse AChR, and transient heterologous expression in appropriate cellular systems. Comparison with the responses obtained with the wild-type subunits in vitro will enable correlations to be established between the clinical and the molecular levels.

Transgenic mice expressing the SCCMS mutants will be used with increasing frequency to learn about the pathogenesis of the degenerative myopathological alterations occurring in these, and other myasthenic syndromes. The combined studies will hopefully lead to identification of targets for potential therapeutic approaches.

Fast Channel Syndrome

Not all congenital syndromes exhibit slower channel kinetics. The fast channel congenital syndrome[38] (see Table 8.2) is a pathology in which patients exhibit very small m.e.p.p.s, but with normal AChR density; there is no endplate AChR deficiency. Single-channel events are infrequent, with diminished channel reopenings during ACh occupancy, and resistance to desensitization by ACh.[38] Two patients were reported, each having two heteroallelic null AChR ε subunit gene mutations: a common εPro121Leu mutation, which defines the pathological phenotype, a signal peptide mutation (ε G-8R) (patient 1), and a glycosylation consensus site mutation (ε Ser143Leu) (patient 2). Studies of the engineered εPro121Leu AChR revealed a significantly decreased rate of AChR channel opening (see Scheme 2 above), little change in affinity of the resting state, R for ACh, but reduced affinity of the open channel and desensitized states (O and D in Scheme 2 above, respectively). Thus the fast channel syndrome is a missense syndrome that causes loss of function, exhibits low affinity for ACh, and requires a heteroallelic null mutation in the same subunit gene to become clinically manifest.[38]

AChR Pathological Findings and Some Therapeutic Prospects in Aging, Alzheimer's and Parkinson's Diseases

Patients suffering from Alzheimer's disease and Parkinson's disease exhibit alteration of several neurotransmitter systems. A marked reduction in the number of neuronal nicotinic AChRs in cerebral cortex has been associated with both pathologies.[52-53] The density of AChRs is also significantly diminished in both diseases.[54] The loss appears to be specific to certain neuronal AChR subtypes.[55]

Cognitive disturbances are associated with aging, Alzheimer's and Parkinson's disease (see the recent monograph by Mann[56] in another volume of the Neuroscience Intelligence Unit series). Initially, the cognitive problems were discussed within the framework of a "cholinergic hypothesis" involving mainly the CNS muscarinic receptors.[57,58] Aging and Alzheimer's disease have also been associated with degeneration of neurons of the ascending cholinergic pathway.[58] The cell bodies of these neurons are located in the basal forebrain, and their axons innervate the amygdala, hippocampus and neocortex. The symptomatology of Parkinson's disease includes extrapyramidal motor dysfunction, cognitive and affective disorders. Brain neuronal AChR has not only been suggested to be affected in these pathologies, but also in human cognitive disorders associated with the normal process of aging.[59] Among these are the age-related deficits in short- and long-term memory, impairment of attention, and delayed reaction time. Among several other observations relevant to the cholinergic system, nicotinic AChR binding sites have been reported to be reduced in number in the cerebral cortex of patients with Alzheimer's disease.[54,60] The high affinity nicotine binding sites are also reduced in cerebral cortex[52] particularly in those areas purported to express mainly the $\alpha4\beta2$ form of the AChR.[61]

In animals, the agonist nicotine facilitates learning and the consolidation of memory.[62] Nicotine can improve task acquisition and has been linked to rapid information processing, arousal, attention and psychological mechanisms. The relationship with age stems from the fact that the decline in AChRs observed in aging produces increased vulnerability to the effects of nicotine blockage. The cholinergic noncompetitive antagonist mecamylamine is a drug acting on CNS AChR and has been used to experimentally reproduce impairment of several cognitive processes in humans.[63-65] It is not known whether this vulnerability can progress towards the degenerative disorders observed in Parkinson's or Alzheimer's diseases. Other therapeutic approaches with nicotinic cholinergic ligands have been directed towards increasing neurotransmitter levels and reducing degeneration of cortical neurons (reviewed in ref. 66). Nicotine, for instance, is reported to ameliorate the attention deficit and improve information processing in Alzheimer's patients.[67-68] It is interesting to note that nicotinic AChRs are expressed in lymphocytes, whose number changes in patients with Alzheimer's or Parkinson's disease.[69] With appropriate probes, therefore, a useful assay could be developed to measure the therapeutic progress of these patients.

The study of the involvement of neuronal nicotinic AChR in learning and memory at the molecular level has received new impetus with the availability of molecular genetic techniques. Knockout experiments using transgenic mice have recently been performed to examine the effect of nicotine on the CNS AChR. This was examined using gene targeting to mutate the $\beta 2$ subunit gene, the most common neuronal AChR subunit expressed in brain; the effect was measured by an experimental learning paradigm.[68] High-affinity nicotine binding sites were found to be absent from brains of mice homozygous for the $\beta 2$ deletion. Furthermore, thalamic neurons from these knockout brains no longer responded to nicotine application. In vivo, the $\beta 2$-deficient animals exhibited an abnormal avoidance response, a test of associative memory. The abnormality paradoxically consisted of a better performance in the behavioral test, in spite of their lack of response to nicotine.

The ligands epibantidine and ABT-418 have enabled the identification of selective losses of $\alpha 4\beta 2$ subtype of neuronal AChR in postmortem temporal cortex of Alzheimer's patients,[55] leading these authors to suggest that this AChR subtype may be the most vulnerable to Alzheimer's disease.

Maelicke and Albuquerque[66] have discussed the pitfalls of using cholinesterase inhibitors to ameliorate the symptoms in Alzheimer's disease. They pointed out that the desensitization phenomenon that supersedes activation of CNS AChRs can be circumvented using a novel class of nicotinic ligands, which potentiate the response of AChRs to the natural neurotransmitter acetylcholine by acting from an allosteric site. They term these ligands positive allosteric modulators, since they allosterically potentiate the effect of classical agonists. These allosteric ligands act themselves as noncompetitive agonists. The plant alkaloid physostigmine, and the compounds galanthamine and codeine, relatively lipophilic compounds, are representative examples of this class of ligands. ABT-418 is also currently under investigation as a possible therapeutic agent for Alzheimer's dementia.

Neurotoxic Substances and Loss of Neuronal AChR in Dementia and Other Neurological Disorders

The prevalence of certain neurological diseases in industrialized countries, and the differences in the characteristics of a disease between urban and rural areas throughout the world has drawn the attention of epidemiologists, as is the case with, for instance, Parkinson's disease. In particular, there appears to be a correlation between the early onset of this disease in industrialized countries, and also between increased risk and a history of rural residence. The demographic factor may have a more precise environmental origin, since increased risk of Parkinson's disease has been correlated with the use of pesticides, herbicides, and industrial chemical exposure. Whereas many commonly used pesticides are not accumulated in the human body, organochlorine pesticides are. Among these are the widely used 2,2-bis(*p*-chlorophenyl)-1,1,1-trichloroethane (DDT) and Lindane, which can be found in adipose tissue of humans exposed to these lipid-soluble pesticides. Although their use was banned in the United States in 1972, they are still used in a number of countries. DDT, its metabolites and Dieldrin have been found in adipose tissue of humans from several parts of the world. pp-DDT has been found in brains of humans with Alzheimer's and Parkinson's diseases. Industrial exposure to manganese, or to the manganese-containing fungicide manganese ethylene-*bis*-(dithiocarbamate) ("Maneb"), and to the fumigant carbon disulfide has also been correlated with Parkinson's disease (ref. 70 and references therein).

Anxiety, Schizophrenia and Neuronal AChR

Schizophrenia is another pathology probably affecting a large number of neurotransmitter systems, nicotinic AChRs included.[71-72] Nicotine ameliorates two of the psychiatric signs found in schizophrenia: the auditory sensory gating deficit and the erratic smooth pursuit eye movements (ref. 72 and references therein). Repeated auditory stimuli normally evoke responses in brain, but these cannot be adequately processed by schizophrenic patients, who fail to habituate to the stimuli. This inability may be related to the distractibility and hypervigilance of psychotic patients. In humans the hippocampus has been implicated as one of the sources for evoked potential that fail to habituate in schizophrenic patients.[73] The neuronal AChR in hippocampus may be involved in the inhibition of the auditory response that is defective in the schizophrenic patients.[74] Cholinergic activation of inhibitory interneurons, primarily in the dentate gyrus and the CA3 region of Ammon's horn would be needed for habituation to the auditory stimulus. The interneurons in these hippocampal regions possess α-bungarotoxin binding sites, and activation of these by cholinergic stimuli would increase inhibitory synaptic input to pyramidal neurons, and thereby diminish the responsiveness of these latter neurons to sensory auditory stimulation. The observation that the number of AChRs is diminished in the hippocampal CA3 region of Ammon's horn and the dentate gyrus lends support to the hypothesis that nicotinic AChR deficit is involved in the clinical manifestation of the auditory functional defect observed in schizophrenics.[72]

Antimuscarinic agents such as scopolamine or some nicotinic cholinergic drugs like mecamylamine do not block the habituation of the auditory evoked responses that are deficient in schizophrenic patients, whereas the competitive nicotinic antagonists α-bungarotoxin and curare do.[74] Nicotine transiently normalizes the evoked potential gating.[75] Schizophrenic patients who are regular smokers also exhibit normal responses to auditory stimuli[76] and also the other pathognomonic signs in some forms of schizophrenia, e.g., disordered smooth pursuit eye movements.[77]

The involvement of the AChR in anxiety is attested by the anxiolytic-like effects reported for nicotine and related compounds in humans and in animal models. ABT-418, an isoxazole isostere of nicotine, is a potent and stereoselective cholinergic ligand for neuronal AChR, as is lobeline. However, nicotine stimulates dopamine release, whereas lobeline does not, an observation which may explain the differences in their behavioral effects (reviewed in ref. 78). The anxiolytic effect of ABT-418 can be blocked by the antagonist mecamylamine, suggesting that the former ligand may act as an AChR agonist.

Involvement of the α4 Neuronal AChR Subunit in Some Forms of Epilepsies

The epilepsies are a phenotypically and genetically heterogeneous group of neurological disorders whose clinically most notorious sign is convulsions. Seizures constitute an alteration of neuronal circuitry that is a network property resulting in intermittent, synchronized bursting of neurons interspersed with periods of calm.

Seizures are, however, only one manifestation of more diffuse brain disorders. Historically, one of the most lasting classifications of the epileptic syndromes is that referring to the clinical and electroencephalographic signatures based on the neuroanatomical region(s) involved, thus giving rise to "generalized" and "partial" (or "focal") forms. Inheritance of the multifactorial type appears to be an important component of epileptic disorders. Although the molecular basis for the common idiopathic epilepsy remains unknown, specific genes have been identified in a few cases involving seizures as part of more diffuse brain pathologies. In two instances these gene alterations correspond to mutations in ion channels. One is the mutation in the α_{1A} gene of the voltage-sensitive Ca^{2+} channel[79] in a generalized, tottering seizure syndrome observed in mutant mice, termed "absence epilepsia" (equivalent to the "petit mal" in humans); the other two refer to AChR mutants, as discussed below.

Indeed, the possible involvement of the neuronal AChR has been postulated in some cases of partial and generalized epilepsies by mapping the locations of the corresponding pathological genes. The first case refers to a particular type of partial epilepsy affecting children, namely the autosomal dominant nocturnal frontal lobe epilepsy (ADNFLE).[40] This is a partial epilepsy causing frequent, violent, brief nocturnal seizures which occur almost exclusively during sleep or drowsiness, usually beginning in childhood. The gene for ADNFLE maps to chromosome 20q13.2-q13.3,[40] thus discarding some mouse models on the putative localization of partial epilepsy genes. Interestingly, the α4 neuronal AChR subunit also maps to the same region of 20q, between markers D20S20 and D20S24.[81] When family members of patients suffering from ADNFLE within a large Australian pedigree were screened for mutations within the gene coding for the neuronal AChR α4 subunit, Steinlein et al[40] found a missense mutation that replaces serine with phenylalanine at codon 248, a strongly conserved amino acid residue in the M2 transmembrane domain. In the peripheral, muscle type AChR, this region of the receptor is highly conserved in all members of the ligand-gated ion channels,[82] and contributes to the wall of the ion channel proper (see ref. 83 and chapter 5 by Ortells et al). Steinlein et al[40] discarded the possibility that the α4 Ser248Phe mutation is a benign polymorphism, especially in view of its strong linkage with ADNFLE. In addition, Ser248 is highly conserved among AChR α subunits in humans, other vertebrates and invertebrates. Interestingly, Ser248, lying at about the mid-region (6' position) of the M2 transmembrane segment, has been implicated in the binding of the noncompetitive antagonists chlorpromazine and phencyclidine (reviewed in ref. 83). Examples of the involvement of Ser248 (6') in AChR conductance have been reported. Thus, site-directed mutagenesis of Ser248 to valine or tyrosine decreased Na^+ and K^+ conductance; substitution of Ser248 for alanine altered the dissociation rate of channel blockers.[83]

The relationship between the point mutation observed in ADNFLE patients and the clinical signs of the disease raises obvious interest. When neuronal $\alpha 4\beta 2$ wild-type AChR and a mutant with a Ser248Phe substitution were heterologously expressed in *Xenopus* oocytes,[41,42] faster desensitization rates and slower recovery from the desensitized state were observed in the mutant receptor. The relationship of these phenomena with the clinical phenotype is not apparent. Weiland et al[41] suggested that the mutation causes seizures by diminishing AChR activity upon conversion into the desensitized state, unresponsive to ACh. The occurrence of seizures almost exclusively at night in ADNFLE led Steinlein et al[40] to speculate about the involvement of cholinergic neurons affecting sleep and arousal at thalamic and cortical levels.[84] Figl et al[42] suggested that the enhanced desensitization rate of the mutant AChR in ADNFLE may be associated with a decline in nicotinic response during high-frequency AChR activity and a decrease in cortical feedback inhibition produced by ACh-induced GABA release.

Epileptic patients under long-term anticonvulsant therapy exhibit some abnormalities of AChR functionality associated in some cases with changes in AChR number.[85] Patients undergoing chronic anticonvulsant therapy have also been examined for their response to succinylcholine during anesthesia. A hypersensitivity to this cholinergic agent was found. Since the abnormal response to succinylcholine was also linked to abnormalities of the ryanodine channel/Ca^{2+} ATPase, the possibility was suggested that such phenomena could be attributed, at least in part, to AChR upregulation.[80]

The gene for ADNFLE maps to chromosome 20q13.2-q13.3 in most, but not all cases of typical forms of this disease.[80] Since additional familial partial epilepsies have been recognized[87] other members of the AChR family might also be involved. The locations of neuronal AChR subunits other than $\alpha 4$ are now being screened as candidate regions for linkage in these non-$\alpha 4$ families. If mutations were to be found in other forms of epilepsy, this could signify an important step towards identifying the clinical phenotype and understanding the etiology of these important human disorders, which affect about 2% of the population.

Genetic studies have also pointed to the involvement of the α4 AChR subunit in other generalized forms of the epilepsies. Among these are the so-called benign familial neonatal convulsions (*EBN1*). In this form of the disease, neonatal convulsions arise spontaneously within a few weeks of birth, and remit spontaneously within a few months. Like ADNFLE, *EBN1* is inherited as a single locus trait.[88] A nonsense mutation in gene coding for the α4 AChR subunit (CHRNA4) cosegregates with the 20q-chromosome.[43-44] Interestingly, the marker D20S19 has been localized very close to markers D20S20 and D20S24 mentioned above.[89] The former marker has been associated with *EBN1*,[88] low-voltage EEG,[90] and ADNFLE.[80] Thus, the possibility arises that phenotypically different forms of epilepsies result from different mutations in the same or related genes. Furthermore, the localization of the α4 AChR gene and the epilepsy genes to similar regions of the 20q chromosome[80] open interesting possibilities for the exploration of other genes with regional cortical expression and/or the vulnerability of certain regions of the cerebral cortex, in addition to deciphering the problem of heterogeneity in idiopathic epilepsies, and identifying "epilepsy susceptibility genes" by positional cloning.

Gilles de la Tourette's Syndrome

Children between five and ten years of age are most prone to have tics and habit spasms. When purposive coordinated movements serving a function are incessantly repeated when uncalled for they may eventually become habit; in more severe cases may progress to become stereotyped, multiple and convulsive. The Gilles de la Tourette syndrome is one such extreme form of movement disorder. The clinical symptoms of patients suffering from Tourette's syndrome are ameliorated by the combined action of neuroleptics like haloperidol and nicotine patches.[91-94] There is no unambiguous demonstration, however, of the involvement of neuronal nicotinic AChRs in CNS fast synaptic transmission particularly in this disease. In rats, nicotine potentiates the catalepsy produced by the dopamine antagonist haloperidol, a paradoxical observation, since nicotine increases dopamine release. More clinical research is needed to evaluate the potential therapeutic effects of nicotine in patients with Tourette's syndrome.

Pain

Antinociception is another property of nicotine and related compounds like ABT-418. Nicotine effects on pain can be attenuated by mecamylamine.[78] The mechanism of analgesia is not known, but it may involve AChRs located either in subcortical areas and implicated in pain regulation via descending pain inhibitory pathways or in the spinal cord. The alkaloid epibatidine, another cholinergic ligand, has an analgesic effect 80-fold more potent than that of morphine.[78] Epibantidine and synthetic analogs are potent selective agonists of the neuronal AChR.

Stress

Various CNS neurotransmitter receptors appear to be altered by stress: β and α2 adrenergic receptors are downregulated, whereas muscarinic AChR are upregulated.[95-96] It has been found that nicotinic CNS AChRs are also modified in stress induced by chronic immobilization of the test animal. Thus, immobilization sustained for 2 hours a day, for a period of two weeks resulted in downregulation of nicotinic AChRs in rat cerebral cortex and midbrain.[97] Chronic nicotine treatment had the opposite effect, abolishing the stress-induced AChR downregulation.

Onchocerciasis

The potentiality of neurotransmitter receptor molecules as targets of therapeutic approaches or as therapeutic agents themselves is only recently beginning to be recognized. One interesting case can be found in the involvement of the AChR in parasitic diseases. Onchocerciasis, a cutaneous filariasis afflicting 18-20 million individuals in Africa, the southern part of North America, South America, and Yemen, is caused by *Onchocerca volvulus*, a nematode parasite of the superfamily Filarioidea. Three million of the affected individuals develop "river blindness" (so-called because the infection is transmitted by flies of the genus *Simulium* which breed along fast-moving rivers), consisting of a keratitis, iridocyclitis and, less commonly, a corioretinitis which can be directly attributed to lesions derived from the parasitic disease.

The most commonly used chemotherapy for onchocerciasis involves the administration of the drug Ivermectin, which is effective against the larval, but not the adult-form of the worm. Given the fact that the AChR of nematodes is the target of other antihelmintic drugs,[98-99] it has been suggested that new drugs could be developed targeting the AChR for the treatment of onchocerciasis.[100] Similarly, the recent cloning of an α-like AChR subunit from the parasitic nematode *Ascaris suum* and the knockout model produced[101] may help to explore new therapeutic strategies against the various stages of these diseases and to elucidate the molecular basis of antiparasitic drug-receptor interactions.

Future Perspectives

We can anticipate the discovery of further acquired and genetic pathologies affecting the AChR in coming years. Several current neurological findings with no clear etiology are likely to find a physiopathological explanation at the molecular level in the light of these developments. Still unexplained clinical findings may also be fathomed when the mechanism of action of various cholinergic compounds on distinct AChR subtypes becomes clearer. This is the case with, for instance, some postoperative syndromes, possibly involving anesthetic or anesthesia-related neuromuscular impairment. Thus, some forms of channel pathologies like the one observed in hyperkalemic cardiac arrest after succinylcholine administration[102] will eventually be characterized at a molecular level in patients suffering from prolonged immobilization, severe burns or trauma, radiation injuries, muscle trauma, or upper motor neuron injuries. The origin of the increase in AChR number after burns at sites distant from the area of injury is still unknown, although the phenomenon appears to differ from that of denervation hypersensitivity.[103]

Single-gene disruption whole animal mutants will be increasingly used to test null-phenotype of altered genes. Knockout mice models will play an important role in the discovery of new AChR pathologies, as well as those involving other nonreceptor proteins present in the cholinergic synapse, such as RAPsyn, agrin, utrophin or laminin (refs. 6, 104-105 and references therein).

In vivo studies of neuronal AChR in the human brain are likely to be tackled in the immediate future if adequate tracer compounds for PET and SPECT noninvasive imaging are developed. This, in combination with the development of selective ligands to dissect the role of AChR subtypes in mediating behavioral effects, will probably result in novel approaches of therapeutic relevance in some CNS neurological diseases involving the AChR and, in particular, in psychiatric disorders.

Drug development will certainly be a major focus of future development. Improved design of drugs with selectivity for different AChR subtypes will probably receive a major impetus. Advances in Parkinson's disease may arise from a better understanding of the complex relationship between the involvement of the cholinergic and dopaminergic systems in extrapyramidal motor dysfunction, and the cognitive and affective symptoms observed in this disease. The major target of therapeutic approaches at present are the motor disorders. Progress in the management of the other two deficits may follow from development of appropriate ligands based on this knowledge.

In schizophrenia, adverse side effects of anti-psychotic medication addressed to the hyperactivity symptoms may be ameliorated by concomitant therapy targeted to the cholinergic system, i.e., by the use of nicotine patches. A better comprehension of the putative involvement of different AChR subtypes in schizophrenia is crucial to a rational therapeutic strategy in this very complex pathology.

On the front-line of genetic diseases, new pathologies will be described and characterized at the molecular level. The discovery of alternative splicing of specific AChR subtypes, as is observed with other ligand-gated ion channels, is a likely possibility. A major breakthrough can be expected in the discovery of new diseases, and possibly animal models thereof, by comparative screening of large databases of human and animal genomes. This is based on the fact that, in spite of the evolutionary distance between *Homo sapiens* and other species like yeast or *Drosophila* (see chapter 2 in this volume), several genes have been found to be conserved among such species and to perform similar or related functions. The eukaryotic yeast genome is complete, and several nervous system-relevant genomes, such as the paradigm for neural development, *Caenorhabditis elegans*, and that of *Drosophila*, are quite advanced. These genomes are small, gene-rich, and intron-poor in comparison to the human genome, which has less than 4% protein coding regions, and is thus effectively equivalent to normalized cDNA libraries (cf. ref. 106). If a given gene is mapped in mouse or rat, then the location of the human gene homologue can often be predicted on the human map, because genes occur in a conserved manner from rodents to humans. This property is called "synteny."[107]

The recognition of rodent-human gene homologues has permitted the development of embryonic stem cell knockout mice. A recent development known as "conditional gene knockout" also permits switching gene expression on and off and replacement of wild-type genes with mutant alleles. The latter is particularly suited to generating mutations instead of gene deletions, and producing animal models of human disease (for example see ref. 108). A mutation in mice of a gene homologous to a human gene does not necessarily lead to the same phenotype in the animal, however. One mutation can have remarkably different phenotypes when expressed in different genetic backgrounds. This is due to the occurrence of different alleles at modifying loci in different mouse strains.[109] A little help from the experimentalist may not hurt: altering metabolic pathways in mice to make them more similar to man appears to be a valid strategy to produce the correct animal model, and crossing a given mouse with several different inbred strains for at least two generations before again breeding homozygotes may also improve the model.[109] Inserting the right mutation by "transgenesis" is another approach that has been applied to study Alzheimer's disease in transgenic mice.[110]

cDNA clones from plasmid DNA libraries have been selected at random and several hundred bases sequenced from both ends. These short sequences are called "expressed sequence tags" (ESTs). The position of a gene or DNA marker on the physical chromosome map is called a sequence tagged site, or STS. On the basis of such knowledge, it has recently been possible to identify and map 66 human cDNAs (ESTs) homologous to mutant phenotypes of *Drosophila* genes by searching through the EST database.[111] Comparative genome screening of large sequence databases of human and other species of known genomic maps and phenotypes will certainly become an invaluable tool to identify the animal complement of human genes and new members of gene families, to discover novel phenotypes of medical interest, and to characterize mutations of the AChR and other ligand-gated channels leading to heritable diseases.

In fact, these combined approaches have already permitted the identification of insect remnants in the human genome. Thus, region p11.2-p12 in human chromosome 17 was found to contain two copies of one such insect remnant, the so-called *mariner*-like transposable element (MLE) separated by about 1.5 megabases. MLE could serve as a hotspot for the initiation of homologous recombination, and the human genome, as well as that of other primates, could contain a large number of such elements. Within this region in the human chromosome is the gene for peripheral myelin protein 22. The duplication and deletion products result in relevant neurological syndromes: duplication gives rise to the Charcot-Marie-Tooth disease type 1A (CMT1A) and the deletion syndrome is hereditary neuropathy with liability to pressure palsies (HNPP). In *Drosophila* (*mauritania*), this coincides with a recombinant hotspot.[112]

Integration of the genes into the transcription map of the human genome will also contribute to categorizing genomic pathologies by using the "positional candidate approach." Gerhold and Caskey[106] have reviewed this approach, which is dramatically speeding up with the advent of EST databases. They identify four steps in such strategy: First, clinical information and cDNA samples are collected from afflicted patients and their relatives. Second, pedigree linkage analysis is undertaken to identify the approximate position of the affected gene by typing polymorphic markers over the entire genome in the collected cDNA samples. Polymorphic markers are short chromosomal DNA sequences. Pedigree linkage analysis is based on the coinheritance of one or more of such markers with the disease, since coinheritance is indicative of proximity of the affected gene with the marker. Pedigree linkage analysis can pin down the search for the disease gene to within a region of 1-10 million base pairs. Sequencing of random, small DNA fragments of the region ("shotgun libraries") is then performed: this procedure is called "sequence skimming." This enables, in turn, the comparison of genomic sequences with EST sequences in the databases. Finally, the cause-effect relationship between the gene and the disease is established by identifying the mutation in a candidate gene that is specifically present in the affected patient—and carriers—but not in normal individuals. As summarized by Gerhold and Caskey,[106] the strategy first identifies sequence polymorphisms close enough to the disease gene to be coinherited with the disease, and then uses such polymorphisms as markers to close in on the mutation that actually causes or contributes to the disease.

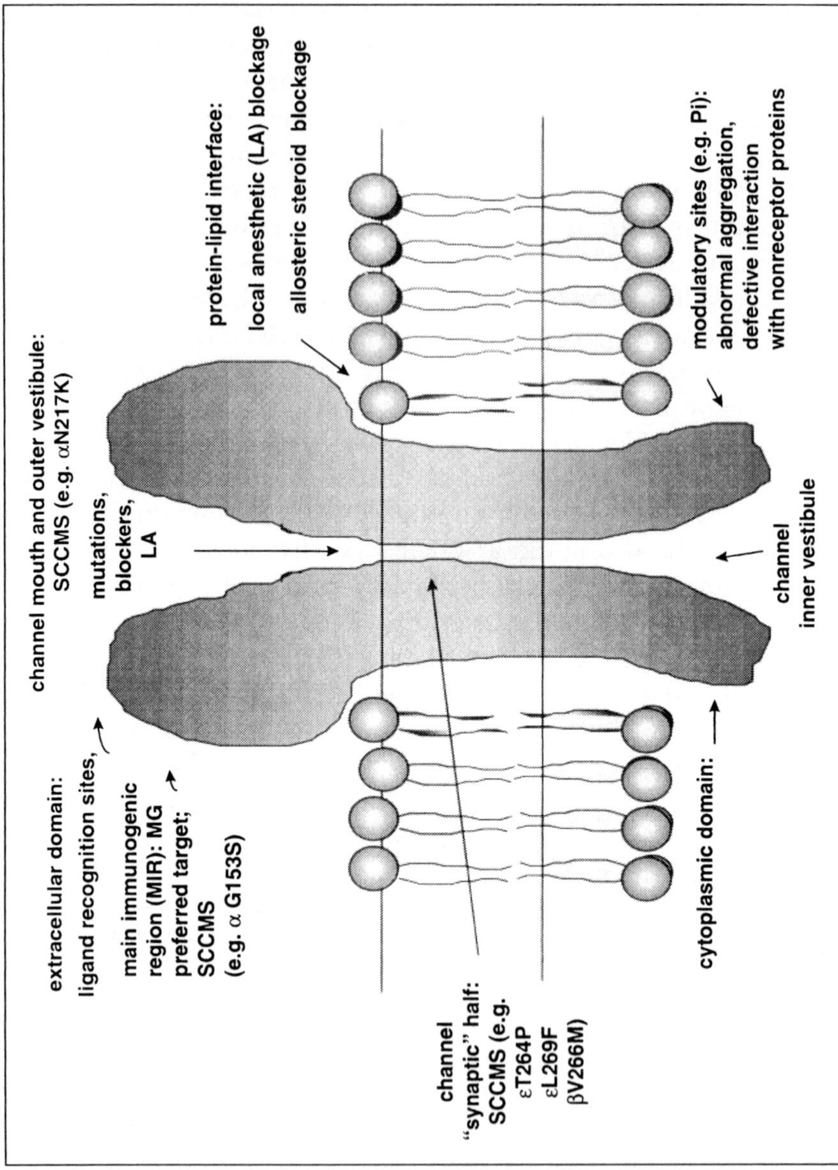

Fig. 8.3. Some possible targets of direct and indirect AChR pathologies. Possible loci of pharmacological agents causing functional alterations or modulation of receptor function are indicated in the scheme. Reprinted and modified with permission from Neurochem Res 1997; 22:391-400.

Concluding Remarks

I have briefly reviewed several pathological conditions affecting the AChR macromolecule. The two essential functions performed by this neurotransmitter receptor, namely ligand (agonist) recognition and channel gating, can be the target of congenital or acquired disease. Thus, abnormal interaction of the AChR with its natural neurotransmitter ACh, or anomalous kinetics of channel opening-closing resulting from mutations in the pore region proper and, interestingly enough, in regions other than the pore, can lead to defective neuromuscular transmission. Less conspicuous modifications of the receptor-agonist interaction, and hence more subtle alterations of AChR function, can result from pharmacological effects exerted by ligands (e.g., local anesthetics, steroids) at the protein-lipid interface,[113-114] or by partial occlusion of the channel by blocking drugs. Figure 8.3 diagramatically summarizes these possible targets of disease and pharmacological regulation of the AChR.

References

1. Bryant SH. Ion channels as targets for genetic disease. In: Sperelakis N, ed. Cell Physiology Source Book. Academic Press, San Diego, 1995:413-427.
2. Rojas CV. Ion channels and human genetic diseases. News Physiol Sci 1996; 11:36-42.
3. Conti-Fine BM, Protti MP, Bellone M, Howard JF. Myasthenia Gravis: The Immunobiology of an Autoimmune Disease. Neuroscience Intelligence Unit, Georgetown, TX: Landes Bioscience, 1997:230.
4. Lindstrom J. Neuronal nicotinic acetylcholine receptors. In: Narahashi T, ed. Ion Channels, vol. 4, Plenum Press, New York, 1996:377-450.
5. Barrantes FJ. The acetylcholine receptor ligand-gated channel as a molecular target of disease and therapeutic agents. Neurochem Res 1997; 22:391-400.
6. Vincent A, Newland C, Croxen R, Beeson D. Genes at the junction-candidates for congenital myasthenic syndromes. Trends in Neurosci 1997; 20:15-22.
7. Milone M, Wang H-L, Ohno K, Fukudome T et al. Slow channel maysthenic syndrome caused by enhanced activation, desensitization, and agonist binding affinity due to mutation in the M2 domain of the acetylcholine receptor α subunit. J Neurosc 1997; in press.
8. Croxen, R, Newland C, Beeson D et al. Mutations in different functional domains of the human muscle acetylcholine receptor α-subunit in patients with the slow-channel congenital myasthenic syndrome. Human Molec Gen 1997; 6:767-774.

9. Maslinski W, Laskowska-Bozek H, Ryzewski J. Nicotinic receptors of rat lymphocytes during adjuvant polyarthritis. J Neurosci Res 1992; 31:336-340.
10. Maneckjee R, Minna JD. Opiod and nicotine receptors affect growth regulation of human lung cancer cell lines. Proc Natl Acad Sci USA 1990; 87:3294-3298.
11. Beroukhim R, Unwin N. Three-dimensional location of the main immunogenic region of the acetylcholine receptor. Neuron 1995; 15:323-331.
12. Davis MM, Bjorkman PJ. T cell antigen receptor cells and T cell recognition. Nature 1988; 334:395-402.
13. Lindstrom J, Shelton F, Fuji Y. Myasthenia gravis. Adv Immunol 1988; 42:233-284.
14. Manfredi AA, Bellone M, Protti MP et al. Molecular mimicry among human autoantigens. Immunol Today 1991; 12:46-47.
15. Engel AG. Myasthenia gravis and myasthenic syndromes. Ann Neurol 1984; 16:519.
16. Vincent A, Newsom-Davis J, Wray D et al. Clinical and experimental observations in patients with congenital myasthenic syndromes. In: Penn AS, Richman DP, Ruff RL, Lennon VA, eds. Myasthenia gravis and related disorders: Experimental and clinical aspects. Ann New York Acad Sci, 1993; 681:451-460.
17. Vincent A, Newsom-Davis J. Acetylcholine receptor antibody as a diagnostic test for myasthenia gravis: 153 validated cases and 2967 diagnostic assays. J Neurol Neurosurg Psychiatry 1985; 47:1246-1252.
18. Penn AS, Richman DP, Ruff RL, Lennon V, eds. Myasthenia gravis and related disorders. Ann N Y Acad Sci, 1993:681.
19. Conti-Tronconi BM, McLane KE, Raftery MA et al. The nicotinic acetylcholine receptor: Structure and autoimmune pathology. Crit Rev Biochem Mol Biol 1994; 29:69-123.
20. Okumura S, McIntosh K, Drachman DB. Oral administration of acetylcholine receptor: effects on experimental myasthenia gravis. Annals of Neurol 1994; 36:704-713.
21. Berta E, Confalonieri P, Simoncini O et al. Removal of anti-acetylcholine receptor antibodies by protein A-immunoadsorption in myasthenia gravis. Internat J Artif Organs 1994; 17:603-608.
22. Araga S, LeBoeuf RD, Blalock JE. Prevention of experimental autoimmune myasthenia gravis by manipulation of the immune network with a complementary peptide for the acetylcholine receptor. Proc Natl Acad Sci USA 1993; 90:8747-8751.
23. Leigh R, Zee D. The Neurology of Eye Movements. Philadelphia: F.A. Davis, 1983.
24. Salpeter MM. Vertebrate neuromuscular junctions: general morphology, molecular organization, and functional consequences. In: Salpeter MM, ed. The Vertebrate Neuromuscular Junction. New York: Alan R. Liss, 1987:1-54.

25. Kaminski HJ, Maas E, Spiegel P et al. Why are eye muscles frequently involved in myasthenia gravis? Neurology 1990; 40:1663-1669.
26. Schuetze S, Role L. Developmental regulation of nicotinic acetylcholine receptors. Annu Rev Neurosci 1987; 10:403-457.
27. Horton RM, Manfredi AA, Conti-Tronconi BM. The 'embryonic' gamma subunit of the nicotinic acetylcholine receptor is expressed in adult extraocular muscle. Neurology 1993; 43:983-986.
28. Kaminski HJ, Kusner LL, Nash KV et al. The γ-subunit of the acetylcholine receptor is not expressed in the levator palpebrae superioris. Neurology 1995; 45:516-518.
29. Bouzat CB, Barrantes FJ. Hydrocortisone and 11-desoxycortisone modify acetylcholine receptor channel gating. NeuroReport 1993a; 4:143-146.
30. Bouzat CB, Barrantes FJ. Acute exposure of nicotinic acetylcholine receptor to the synthetic glucocorticoid dexamethasone alters single-channel gating properties. Molec Neuropharm 1993b; 3:109-116.
31. Bouzat CB, Barrantes FJ. Modulation of muscle nicotinic acetylcholine receptors by the glucocorticoid hydrocortisone. Possible allosteric mechanism of channel blockade. J Biol Chem 1996; 271:25835-25841.
32. Sine SM, Ohno K, Bouzat C et al. Mutation of the acetylcholine receptor α subunit causes a slow-channel myasthenic syndrome by enhancing agonist affinity. Neuron 1995; 15:229-239.
33. Newland C, Croxen R, Beeson A et al. Mutation in the human muscle ACh receptor in congenital myasthenia prolong receptor activation. J Physiol 1996; 495:79.
34. Engel AG, Ohno K, Milone M et al. New mutations in acetylcholine receptor subunit genes reveal heterogeneity in the slow-channel congenital myasthenic syndrome. Human Mol Genetics 1996; 5:1217-1227.
35. Gomez CM, Maselli R, Gammack JT et al. A beta-subunit mutation in the acetylcholine-receptor channel gate causes severe slow-channel syndrome. Ann Neurol 1996b; 39:712-723.
36. Gomez, CM, and Gammack, JT. A leucine-to-phenylalanine substitution in the acetylcholine-receptor ion-channel in a family with the slow-channel syndrome. Neurology 1995; 45:982-985.
37. Ohno K, Hutchinson DO, Milone M et al. Congenital myasthenic syndrome caused by acetylcholine receptor channel openings due to a mutation in the M2 domain of the ε subunit. Proc Natl Acad Sci USA 1995; 92:758-762.
38. Ohno K, Wang H-L, Milone M et al. Congenital myasthenic syndrome caused by decreased agonist binding affinity due to a mutation in the acetylcholine receptor ε subunit. Neuron 1996; 17:157-170.
39. Engel AG, Ohno K, Bouzat C et al. End-plate acetylcholine receptor deficiency due to nonsense mutations in the ε subunit. Ann Neurol 1996b; 40:810-817.

40. Steinlein O, Mulley JC, Proping P et al. A missense mutation in the neuronal acetylcholine receptor α4 subunit is associated with autosomal dominant nocturnal frontal lobe epilepsy. Nature Genetics 1995; 11:201-203.
41. Weiland S, Witzemann V, Villarroel A et al. An amino acid exchange in the second transmembrane segment of a neuronal nicotinic receptor causes partial epilepsy by altering its desensitization kinetics. FEBS Lett 1996; 398:91-96.
42. Figl A, Viseshakul N, Forsayeth J et al. A mutation associated with epilepsy enhances desensitization of the α4β2 neuronal nicotinic receptor. Biophys J 1997; 72:A150.
43. Beck C, Moulard B, Steinlein O et al. A nonsense mutation in the α4 subunit of the nicotinic acetylcholine receptor (α4) cosegregates with 20q-linked benign familial neonatal convulsions (EBN1). Neurobiol Dis 1994; 1:95-99.
44. Schubert S, Laconne F, Lefterov I et al. Towards positional cloning of the locus for benign neonatal epilepsy (EBN1) on chromosome 20. Am J Hum Genet 1994; 55 [Suppl. 3]:A270.
45. Uchitel O, Engel AG, Walls TG et al. Congenital myasthenic syndromes. II. Syndrome attributed to abnormal interaction of acetylcholine with its receptor. Muscle Nerve 1993; 16:1293-1301.
46. Engel AG, Hutchinson DO, Nakano S et al. Myasthenic syndromes attributed to mutations affecting the epsilon subunit of the acetylcholine receptor. Ann NY Acad Sci 1993; 681:496-508.
47. Ohno K, Quiram P, Milone M et al. Congenital myasthenic syndromes due to heteroallelic nonsense/missense mutations in the acetylcholine receptor ε subunit gene: Identification and functional characterization of six new mutations. Human Molec Gen 1997; 6:753-766.
48. Wang, H-L, Auerbach A, Bren N et al. Mutation in the M1 domain of the acetylcholine receptor α subunit decreases the rate of agonist dissociation. J Gen Physiol 1997; in press.
49. Unwin N. Nicotinic acetylcholine receptor at 9 Å resolution. J Mol Biol 1993; 229:1101-1124.
50. Gomez CM, Bhattacharyya BB, Charnet P et al. Transgenic mouse model of the slow-shannel syndrome. Muscle and Nerve 1996a; 19:79-87.
51. Treinin M, Chalfie M. A mutated acetylcholine receptor subunit causes neuronal degeneration in *C. elegans*. Neuron 1995; 14:871-877.
52. Flynn D, Mash D. Characterization of 1-[^{3}H]nicotine binding in human cerebral cortex: Comparison between Alzheimer's disease and the normal. J Neurochem 1986; 47:1948-1954.
53. Giacobini E. Cholinergic receptors in human brain: Effects of aging and Alzheimer's disease. J Neurosci Res 1990; 27:548-560.
54. Aubert I, Araujo DM, Cécyre D et al. Comparatve alterations of nicotine and muscarinic binding sites in Alzheimer's and Parkinson's diseases. J Neurochem 1992; 58:529-541.

55. Warpman U, Nordberg A. Epibantidine and ABT 418 reveal selective losses of α4β2 nicotinic receptors in Alzheimer brains. NeuroReport 1995; 6:2419-2423.
56. Mann DMA. Sense and Senility: The Neuropathy of the Aged Human Brain. Neuroscience Intelligence Unit, Georgetown, TX: Austin, 1997:198.
57. Drachman D. Memory and cognitive function in man: Does the cholinergic system have a specific role? Neurology 1977; 27:783-790.
58. Bartus RT, Dean RD, Beer B. The cholinergic hypothesis of geriatric memory dysfunction. Science 1982; 217:408-414.
59. Court JA, Piggott MA, Perry EK et al. Age associated decline in high-affinity nicotine binding in human brain frontal-cortex does not correlate with the change in choline-acetyltransferase activity. Neurosci Res Commun 1992; 10:125-133.
60. London ED, Ball MJ, Waller SB. Nicotinic binding sites in cerebral cortex and hyppocampus in Alzheimer's disease. Neurochem Res 1992:14:745-750.
61. Perry EK, Morris CM, Court JA et al. Alteration in nicotine binding sites in Parkinson's disease, Lewy body dementia and Alzheimer's disease: possible index of early pathology. Neuroscience 1995; 64:385-395.
62. Nelson JM, Goldstein L. Improvement of performance on an attention task with chronic nicotine treatment in rats. Psychopharmacologia 1982; 26:347-360.
63. Newhouse PA, Potter A, Corwin JR et al. Acute nicotinic blockade produces cognitive impairment in normal humans. Psychopharmacology 1992; 108:480-484.
64. Newhouse PA, Potter A, Corwin J et al. Age-related effects of the nicotinic antagonist mecamylamine on cognition and behavior. Neuropsychopharmacology 1994; 10:93-107.
65. Newhouse PA, Potter A, Corwin J. Effects of nicotinic cholinergic agents on cognitive functioning in Alzheimer's and Parkinson's disease. Drug Devel Res 1996; 38:278-289.
66. Maelicke A, Albuquerque EX. Drug Discovery Today 1996; 1:53-59.
67. Jones GMM, Sahakian BJ, Levy R et al. Effects of acute subcutaneous nicotine on attention, information processing and short-term memory in Alzheimer's disease. Psychopharmacology 1992; 108:448-451.
68. Piccioto MR, Zoll M, Léna C et al. Abnormal avoidance learning in mice lacking functinal high-affinity nicotine receptor in the brain. Nature 1995; 374:65-67.
69. Adem A, Norberg A, Bucht G et al. Extraneural cholinergic markers in Alzheimer's and Parkinson's disease. Biol Psychiatry 1986; 10:247-257.
70. Fleming L, Mann JB, Bean J et al. Parkinson's disease and brain levels of organochlorine pesticides. Ann Neurol 1994; 36:100-103.

71. Goff WR, Henderson DC, Amico E. Cigarette smoking in schizophrenia: Relationship to psychopathology and medication side effects. Am J Psychiatry 1992; 149:1189-1194.
72. Freedman R, Hall M, Adler LE et al. Evidence in postmortem brain tissue for decreased numbers of hippocampal nicotinic receptors in schizophrenia. Biol Psychiatry 1995; 38:22-33.
73. Goff WR, Williamson PD, VanGilder JC et al. Neural origins of long latency evoked potentials recorded from the depth and from the cortical surface of the brain in man. Progr Clin Neurophysiol 1980; 7:126-145.
74. Luntz-Leybman V, Bickford P, Freedman R. Cholinergic gating of response to autidory stimuli in rat hippocampus. Brain Res 1992; 587:130-136.
75. Adler LE, Hoffer LJ, Griffith J et al. Normalization by nicotine of deficient auditory sensory gating in the relatives of schizophrenics. Biol Psychiatry 1992; 32:607-616.
76. Adler LE, Hoffer LJ, Wiser A et al. Cigarette smoking normalizes auditory physiology in schizophrenics. Am J Psychiatry 1993; 150:1856-1861.
77. Klein C, Andersen B. On the influence of smoking upon smooth pursuit eye movements of schizophrenics and normal controls. J Psychophysiol 1991; 5:361-369.
78. Decker MW, Brioni JD, Bannon AW et al. Diversity of neuronal nicotinic acetylcholine receptors: Lessons from behavior and implications for CNS therapeutics. Life Sci 1995; 56:545-570.
79. Fletcher CF, Lutz CM, O'Sullivan TN et al. Absence epilepsy in tottering mutant mice is associated with calcium channel defects. Cell 1996; 87:607-617.
80. Phillips HA, Scheffer IE, Berkovic, SF et al. Localization of a gene for autosomal dominant nocturnal front lobe epilepsy to chromosome 20q13.2. Nature Genet 1995; 10:117-118.
81. Steinlein O, Smigrodzki R, Lindstrom J et al. Refinement of the localization of the gene for neuronal nicotinic acetylcholine receptor $\alpha 4$ subunit (CHRNA4) to human chromosome 20q13.2-q13.3. Genomics 1994; 22:493-495.
82. Ortells MO, Lunt GG. Evolutionary history of the ligand gated ion channel superfamily. Trends in Neurosci 1995; 18:121-127.
83. Bertrand D, Galzi J-L, Devillers-Thiéry A et al. Stratification of the channel domain in neurotransmitter receptors. Curr Opin Cell Biol 1993; 5:688-693.
84. Steriade M, McCormick DA, Sejnowski TJ. Thalamocortical oscillations in the sleeping and aroused brain. Science 1993; 262:679-685.
85. Kim CS, Arnold FJ, Itani MS et al. Decreased sensitivity to metocurine during long-term phenytoin therapy may be attributable to protein binding and acetylcholine receptor changes. Anesthesiology 1993; 77:500-506.

86. Melton AG, Antognini JF, Gronert GA. Prolonged duration of succinylcholine in patients receiving anticonvulsants: evidence for mild upregulation of acetylcholine receptors? Can J Anaesth 1993; 40:939-942.
87. Scheffer IE, Hopkins IJ, Harvey AS et al. New autosomal dominant partial epilepsy syndrome. Ped Neurol 1994; 11:95.
88. Leppert M, Anderson VE, Quattlebaum T et al. Benign familial neonatal convulsions linked to genetic markers on chromosome 20. Nature 1989; 337:647-648.
89. Malafosse A, Leboyer M, Dulac O, Navalet Y, Plouin P, Beck C, Laklou H, Mouchnino G, Grandscene P, Valee L et al. Confirmation of linkage of benign familial neonatal convulsions to D20S19 and D20S20. Hum Genet 1992; 89:54-58.
90. Steinlein O, Anokhin A, Yping M et al. Localization of a gene for a human low-voltage EEG on 20q and genetic heterogeneity. Genomics 1992; 12:69-73.
91. Silver AA, Sandberg PR. Transdermal nicotine patch and potentiation of haloperidol in Tourette's syndrome. Lancet 1993; 342:182.
92. Sandberg PR. Beneficial effects of nicotine in Tourette's syndrome. International Symposium on Nicotine: The Effects of Nicotine on Biological Systems; 1994:II-S39.
93. Silver AA, Shytle R, Philipp M et al. Transdermal nicotine in Tourette's syndrome. In: Clarke PBS, Quik M, Adlkofer F and Thurau K, eds. Effects of nicotine on biological systems II. Basel: Birkhäuser Verlag, 1995:293-299.
94. Arneric SP, William M. Neuronal nicotinic acetylcholine receptors: novel targets for CNS therapeutics. In: Psychopharmacology: The Fourth Generation of Progress. New York: Raven Press, 1995: 1001-1016.
95. Gonzalez AM, Pazos A. Modification of muscarinic acetylcholine receptors in the rat brain following chronic immobilization stress: An autoradiographic study. Eur J Pharmacol 1992; 223:25-31.
96. Takita M, Kigoshi S, Muramatsu I. Effects of bevantonol and hydrochloride on immobilization stress-induced hypertension and central β-adrenoceptors in rats. Pharmacol Biochem Behav 1993; 45:623-627.
97. Takita M, Muramatsu I. Alteration of brain nicotinic receptors induced by immobilization stress and nicotine in rats. Brain Res 1995; 681:190-192.
98. Lewis JA, Wu C-H, Levine JH et al. Levamisole-resistant mutants of the nematode *Caenorhabditis elegans* appear to lack pharmacological acetylcholine receptors. Neuroscience 1980; 5:967-989.
99. Harrow ID, Gration KAF. Mode of action of the antihelmintics morantel, pyrantel, and levamisole on muscle cell membrane of the nematode *Ascaris suum*. Pestic Sci 1985; 16:662-675.

100. Ajuh PM, Egwang TH. Cloning of cDNA encoding a putative nicotinic acetylcholine receptor subunit of the human filarial parasite *Onchocerca volvulus*. Gene 1994; 144:127-129.
101. Brooks HL, Foreman RC, Burke JF et al. Cloning and alpha-like nicotinic acetylcholine receptor subunit from the parasitic nematode *Ascaris suum*. Soc Neurosci 1996: Abstr. 501.10.
102. Gronert GA, Theye RA. Pathophysiology of hyperkalemia induced by succinylcholine. Anesthesiology 1975; 43:89-99.
103. Ward JM, Rosen KM, Martyn JAJ. Acetylcholine receptor subunit mRNA changes in burns are different to that seen after denervation. J Burn Care Rehab 1993; 14:595-601.
104. Noakes PG, Gautam M, Mudd J et al. Aberrant differentiation of neuromuscular junctions in mice lacking s-laminin/laminin β2. Nature 1995; 374:258-262.
105. Gautam, M, Noakes PG, Mudd J et al. Failure of postsynaptic specialization to develop at neuromuscular junction of rapsyn-deficient mice. Nature 1995; 377:232-236.
106. Gerhold D, Caskey CT. It's the genes! EST access to human genome content. BioEssays 1996; 18:973-981.
107. Chung WK, Kehoe LP, Chua M et al. Mapping of the Ob receptor to 1p in a region of nonconserved gene order from mouse and rat to human. Genome Res 1996; 6:431-438.
108. Sands A, Donehower LA, Bradley LA. Gene-targeting and the p53 tumor-suppressor gene. Mutation Res 1994; 307:557-572.
109. Erickson RP. Mouse models of human genetic disease: which mouse is more like human? BioEssays 1996; 18:993-998.
110. Games D, Adams D, Alessandrini R et al. Alzheimer-type neuropathology in transgenic mice overexpressing V717F β-amyloid precursor protein. Nature 1995; 373:523-527.
111. Banfi S, Borsani G, Rossi E et al. Identification and mapping of human cDNAs homologous to *Drosophila* mutant genes through EST database searching. Nature Genetics 1996; 13:167-174.
112. Hartl DL. The most unkindest cut of all. Nature Genetics 1996; 12:227-229.
113. Barrantes FJ. The lipid annulus of the nicotinic acetylcholine receptor as a locus of structural-functional interactions. In: Watts A, ed. Protein-Lipid Interactions. New Comprehensive Biochemistry. Amsterdam: Elsevier, 1993:231-257.
114. Barrantes FJ. Pharmacological sites for some local anesthetic and steroid ligands at the nicotinic acetylcholine receptor-lipid interface. Proc 24th Central European Congress on Anesthesiology. Vienna, Austria. Monduzzi Editore S.p., Bologna, Italia, 1995:487-492.

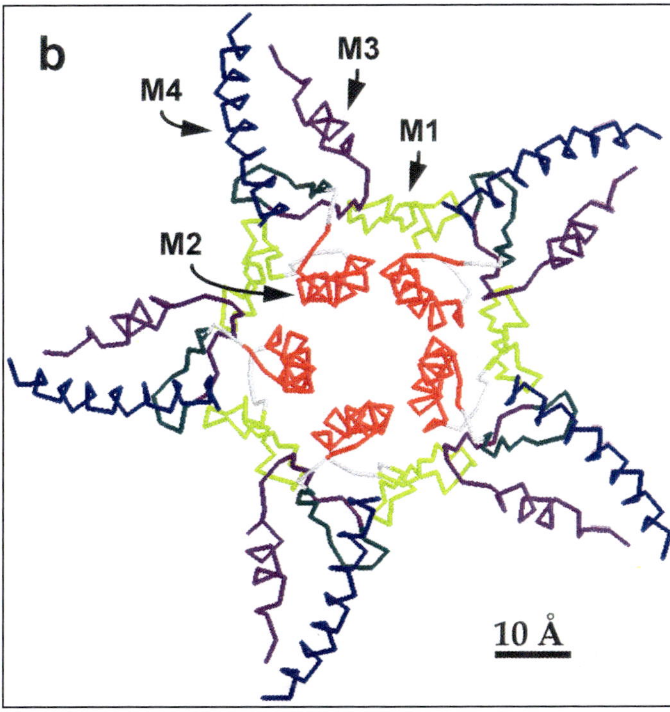

Fig. 5.1. Comparison between Unwin's image[4] (a) and the AChR closed-channel model of ref. 25 (b). Both views are extracellular. Yellow: M1; red: M2, purple: M3; blue: M4; white: connecting loops; green: region previously thought to be in a loop (between M2 and M3) but now part of a β-strand within the membrane in the model. Reprinted with permission from Ortells MO et al, Prot Engng 1996; 9:51-59.

Fig. 5.2. Comparison between Unwin's image[4] (a) and the model (b). Both views are from the side, i.e., as seen from the lipid bilayer. The broken line in (a) represents the boundaries of the bilayer. See Fig. 5.1 for color codes. Reprinted with permission from Ortells MO et al, Prot Engng 1996; 9:51-59.

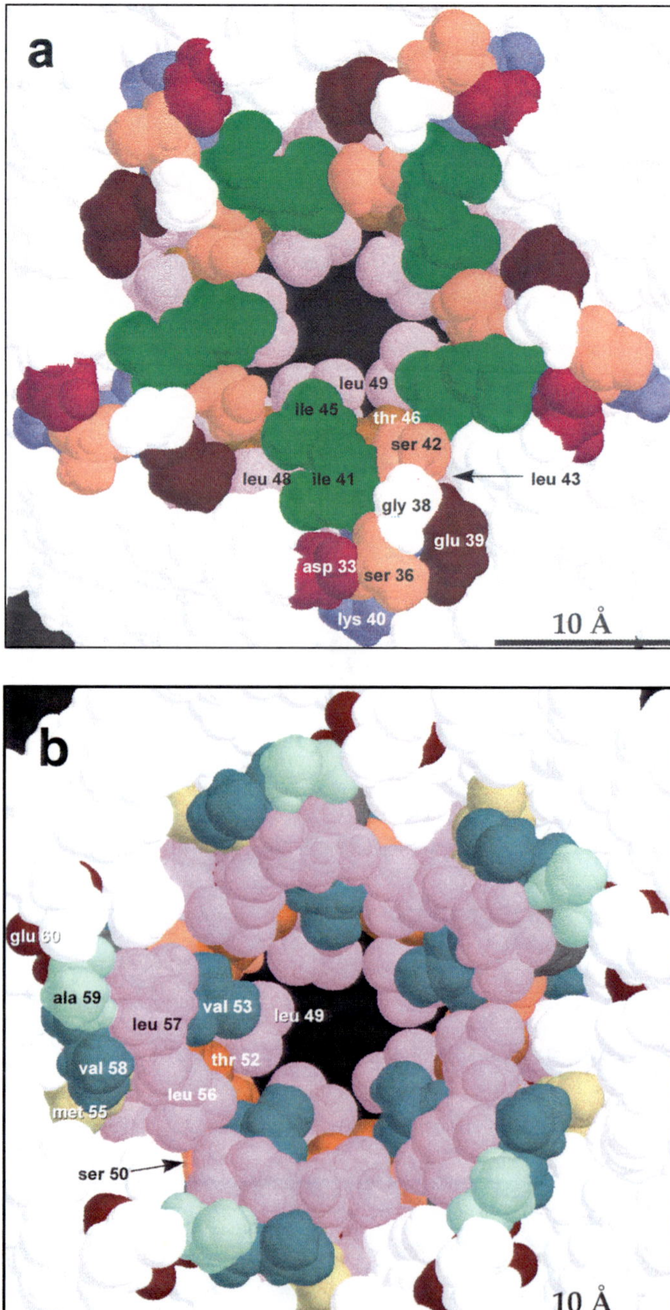

Fig. 5.3. Detailed view of the residues in the AChR ion channel region (a) cytoplasmic view; (b) synaptic view. Reprinted with permission from Ortells MO et al, Prot Engng 1996; 9:51-59.

a

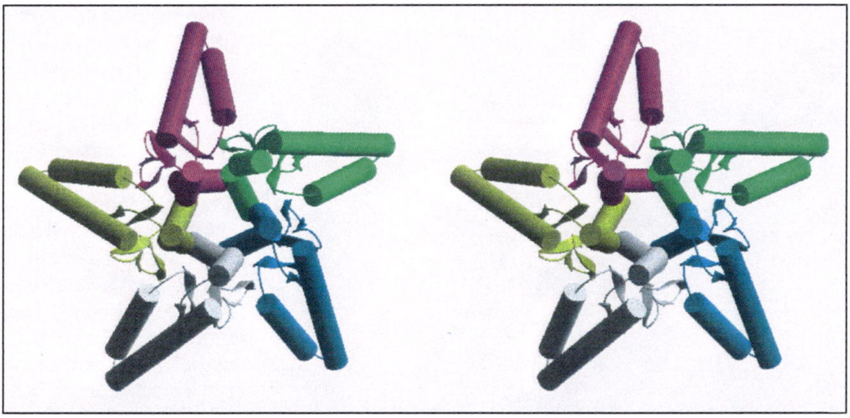

b

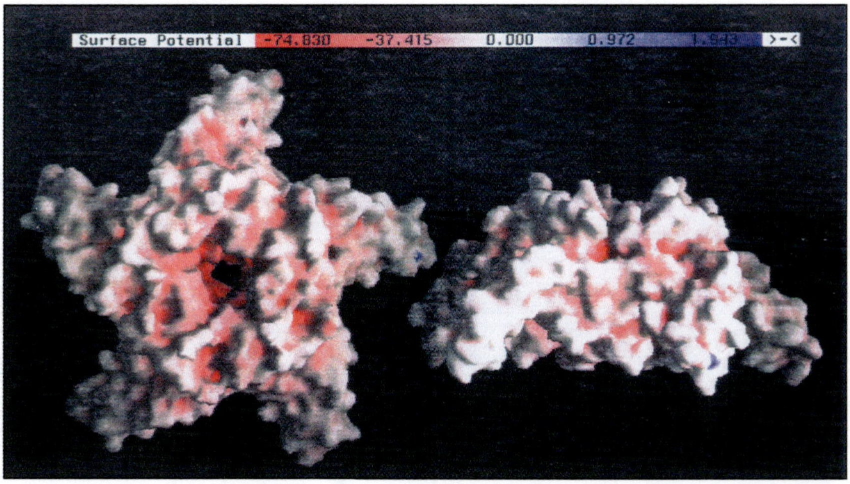

Fig. 5.4. (a) Schematic synaptic view of the whole transmembrane region of the AChR. Each of the five subunits is colored differently. Cylinders are α-helices; flat ribbons are β-strands, and ropes are loops. Generated with the program SETOR (Evans, 1993). (b) Molecular surface generated by the program GRASP, and colored by the electrostatic potential calculated by the program Delphi. Left: synaptic view. Right: lateral (membrane) view. Reprinted with permission from Ortells MO et al.[9]

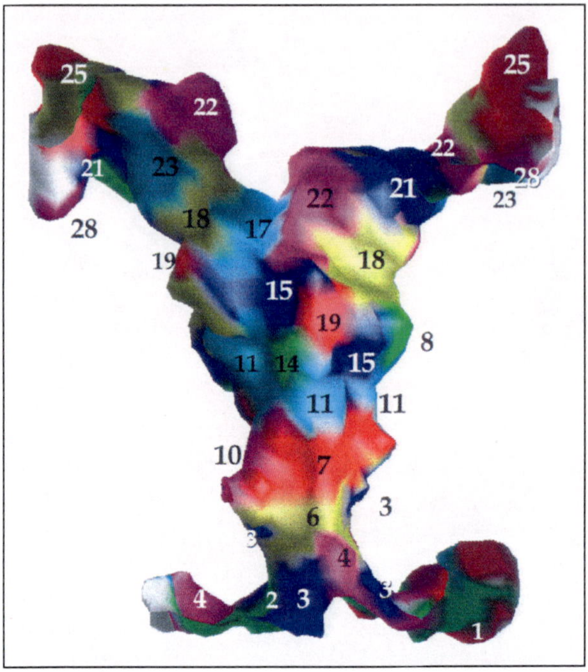

Fig. 5.5. Molecular surface of the ion channel lumen, as viewed from outside the "envelope." Residue numbering 1 to 25 corresponds to that in Table 5.2. Numbers 26, 27 and 28 correspond to the α7 subunit residues Tyr 209 from M1, Met 278 and Ile 279 from M3, respectively. Residue coloring representation is: Red: -5, 2, 8, 14 and +1; Green: -4, 3, 9, 15 and 26; Blue: -3, 4, 10, 16 and 27; Magenta: -2, 5, 21, 17 and 28; Yellow: 1, 7, 13 and 19. Generated by the program GRASP. Reprinted with permission from Ortells MO et al, Prot Engng 1997; (in press).

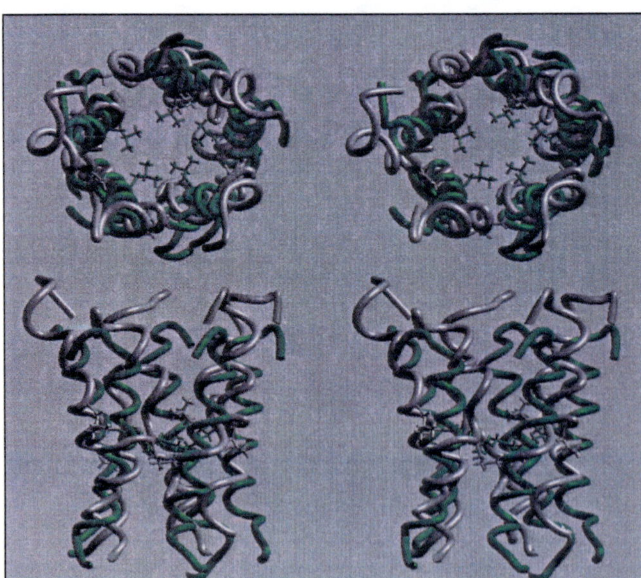

Fig. 5.6. Schematic stereo view of superimposed M2 helices in the closed (green) and open (gray) states (rms value of 6.9). Above: cytoplasmic view; below: lateral view, with the cytoplasmic half of the channel in the upper part of the figure. The side chain of the leucines at position 9 of Table 5.2 are also displayed to show the differences in their position in the open and closed states, respectively. Reprinted with permission from Ortells MO et al.[29]

Index

D

E

F

G

H

I

N

O

P

Q

R

S

T

U

V

W

X

Y